# Spaces of Puppets in Popular Culture

This first book-length exploration of geographical engagement with puppets examines constructions of puppets in contemporary popular British culture and considers the various ways in which puppets and humans (not just puppeteers) are unified in diverse cultural media.

Organised around themes of metaphorical, performative and transformational puppets, the work draws out how puppets are used in diverse cultural media (fiction, music, television, film and theatre), how they are constructed through those uses and to what effect. Both puppets as generalised forms (bodily, relational or ideational) and specific puppet characters (Mr Punch, Pinocchio) are explored. Building upon existing associations between puppets and the grotesque, the volume extends understandings of the puppet by elaborating borderscaping strategies through which puppets are constructed and an alternative perspective on the uncanniness of puppets. Geographically, it unearths distinct puppet spatialities, identifies the socially critical potential of puppets, rescales geo/bio-politics at the interpersonal level and highlights the potential of puppets within posthuman debates about the status of the human.

This work will be of interest to anyone fascinated by puppets, as well as those in fields such as geography, anthropology, cultural and media studies, and those interested in the grotesque, posthumanism and/or non-representational scholarship.

**Janet Banfield** is a college lecturer in Geography at the University of Oxford.

**Routledge Research in Culture, Space and Identity**
Series Editor: Peter Merriman

The *Routledge Research in Culture, Space and Identity Series* offers a forum for original and innovative research within cultural geography and connected fields. Titles within the series are empirically and theoretically informed and explore a range of dynamic and captivating topics. This series provides a forum for cutting-edge research and new theoretical perspectives that reflect the wealth of research currently being undertaken. This series is aimed at upper-level undergraduates, research students and academics, appealing to geographers as well as the broader social sciences, arts and humanities.

For more information about this series, please visit:
www.routledge.com/Routledge-Research-in-Culture-Space-and-Identity/book-series/CSI

# Spaces of Puppets in Popular Culture

## Grotesque Geographies of the Borderscape

**Janet Banfield**

Routledge
Taylor & Francis Group

LONDON AND NEW YORK

First published 2022

by Routledge
4 Park Square, Milton Park, Abingdon, Oxon OX14 4RN

and by Routledge
605 Third Avenue, New York, NY 10158

*Routledge is an imprint of the Taylor & Francis Group, an informa business*

*British Library Cataloguing-in-Publication Data*
A catalogue record for this book is available from the British Library

*Library of Congress Cataloging-in-Publication Data*
A catalog record has been requested for this book

ISBN: 978-1-032-10341-9 (hbk)
ISBN: 978-1-032-10344-0 (pbk)
ISBN: 978-1-003-21486-1 (ebk)

DOI: 10.4324/9781003214861

Typeset in Times
by SPi Technologies India Pvt Ltd (Straive)

From Henson, for Jean: His very special friend.

# Contents

# Tables

# Figures

# About the Author

**Janet Banfield** is a college lecturer in Geography at the University of Oxford. Janet's research and publications draw strongly on Psychology as well as Geography, but her topical research focus lies in Cultural Geography, exploring the creation of space and experience of place through cultural forms and activities from artistic practice to puppetry. From a theoretical perspective, Janet works in 'non-representational' Geography/Psychology, which is concerned both with the work that representations do in the world (i.e., the difference that they make) and with the role of the non-representational (e.g., feelings, sensations, intuition) in the workings of the world. Her academic interests are practice-based and autoethnographic and are concerned as much with methodological development as with the generation of new empirical knowledge and conceptual frameworks.

# Acknowledgements

This book has been ten years in the making, as I first came up with the idea of a geographical exploration of puppets and puppetry shortly before embarking on my doctoral research (into artistic practice). Despite recognising that all the same questions I wanted to pursue in my doctoral research (and more) could be explored through puppets a few short months before starting the doctorate, the puppets had to take a back seat. Puppets, being puppets, though, don't much like being relegated to the back seat, and they would not let me leave them there, so over the past decade, I have been keeping notes on anything I came across that was puppet-related, squirrelling away thoughts, reflections and sentiments on all things puppet, and getting to puppet festivals and performances whenever possible, alongside my doctoral studies and – subsequently – my early academic career. However, to say that *I* have been doing this – in the first person singular – misrepresents both the collective endeavour and the social support upon which this work is founded.

For one thing, Mum and Dad – as ever – have been amazing, both in doing their best to share my (somewhat bonkers) cultural and disciplinary interests and in rooting out data and sources for analysis. While both have identified additional avenues to pursue through their identification of puppet references in whatever they happen to be reading at the time, Mum has been especially and consistently focused in this respect. Mum has been diligently and tenaciously extracting puppet-relevant articles from their daily newspaper and recording a written verbatim log of any references to puppets in the literary fiction that she has read, for at least the past seven years. If I had any allowance or budget to cover 'research assistance' costs, Mum would surely warrant some degree of payment, although I know that financial reasons are not her motivation (not least because she knows that I cannot pay her). Much of the proceeds of Mum's diligent ferreting out of puppet references (both media and fiction-related) has been incorporated into the analyses and discussions contained within this volume, and most of the rest is set aside for an anticipated subsequent volume on the political nature of puppets. Irrespective of where this empirical material ends up fitting within my overall research programme, it is hugely appreciated, and Mum deserves due credit for the foundational role that she has played in enabling me to bring this volume to a close far sooner than would otherwise be the case.

In addition, I extend my thanks to Faye Leerink and Yassar Arafat at Routledge for their editorial and technical expertise, guidance and support, and I am grateful to the anonymous reviewers for their constructive, detailed and helpful suggestions, as well as for the motivational boost that they provided. I also thank all those colleagues and students who have been supportive – or even just mildly tolerant – of my passion for puppets, especially those who have attended workshops I have held on puppet geographies, and those who have accommodated Henson's attendance at social events.

Speaking of Henson (a ginger-furred, sleeve puppet in the form of a primate, named after Brian Henson, creator of The Muppets), I am not entirely sure where he fits with respect to acknowledgements. Henson has clearly brought a lot to my life, both personally and professionally, but he is – after all – an object, not a subject. Is he, though? That is the whole point, isn't it? He has his own character and behaviourisms, despite being a slack-limbed, fabricated, inert thing. He was quite colourist at first, showing a distinct favouritism towards other ginger-toned puppets, and 'aromatist' in that he condemned another puppet (Berta) on the basis that 'she stinks'. Admittedly, having travelled from Prague in a box, the resins, paints and so on, had generated a certain aroma that was released when she came out of the box to join her new 'friends', but Henson took notable exception to this particular scent. To this day, Henson and Berta sit on separate sides of the room, but Henson is now close friends with a blue puppet (Spike) so his colourist – if not his 'aromatist' – days appear (thankfully) to be behind him. Henson also loves what he calls a 'road trip' (travelling between my home and that of my parents) and has clear musical preferences, enjoying his 'head-banging' version of dancing to Dad's on-road musical choices more than almost anything else, except his jocular antics with certain members of the family…

More broadly, I owe thanks to my friends, specifically, Kumiko for accompanying me when we encountered Kermit the Frog as undergraduate geography students and the Bates family for gifting Henson to me for my 40th birthday. I also thank wholeheartedly my maternal aunt – Jean – for being such a great sport in her entertainment of Henson's antics and affection for her over the past few years. While Mum is also a 'beneficiary' of Henson's attentions – especially of the antic variety – it is Jean for whom Henson seems to have special affection, and Jean's unquestioning willingness to go along with Henson's bizarre flow has been invaluable in the evolution of his character. Equally important has been Dad's early reluctance to engage with Henson as a subject rather than an object, causing distinct offence to Henson through Dad's repeated reference to Henson as 'it' rather than 'he' or 'him', to the extent that Henson now identifies Dad as 'It Boy', which is balanced in a gendered way by Henson's acknowledgement of me only as 'The Woman'. While Henson has no direct puppet family of his own, he has certainly had an impact on my family, even though he has had very few opportunities during the pandemic to come out and play. He did, though, make the most of the VE-75 celebrations in 2020 to entertain the neighbours and passing traffic, and I thank Henson for getting me out of the house, for re-igniting my social interest and for energising me so powerfully at that time. It was especially fitting that this occasion involved the Bates family, who had originally gifted Henson to me, so special thanks go to Stu, Charlotte, Steph, Kris and Ethan (and their puppets).

Given this pandemic context, I am especially pleased to be able to bring this work to a conclusion within ten years of conjuring the original idea of a geography of puppets, and – irrespective of the pandemic – I am hugely indebted to everyone named above and to a host of other individuals, who have helped to keep me both grounded and motivated in very trying times. Moreover, I have been surprised and amazed by just how conceptually, affectively and grotesquely productive puppets have proved themselves to be through their geographical analysis, and I am delighted that puppets are now much better positioned to imprint their puppetness on geography, with generative potential for both puppets and geography (and me...). On that note, I am confident that my interest in puppets and puppetness will continue to nurture my own special brand of bonkers in both my private and professional lives, and in the ongoing blending of the two, help me to be both becoming-puppet and my bonkers best.

# Introduction

In October 1994, Kermit the Frog spoke at the Oxford Union, UK, in an event that made national headlines. It is perplexing that a green, felt object can be invited to address an esteemed university's debating society, that such an event can be a source of fascination at the national level, and that the opportunity to hear the words of this fabricated amphibian was such a powerful draw for students who really ought to have better things to do with their time that it was massively over-subscribed. It is a testament to the power of puppets – and specifically this puppet – as a cultural force, and I should know; I was one of those students who should have had something better to do with their time. I was one of the lucky few whose submitted question to the green felt sage was accepted: 'Do you think British children are in any way disadvantaged by the fact that *Sesame Street* is teaching them American spellings?' Kermit's response was that he thought the moral messages of *Sesame Street* were more important than the spelling lessons. It was a magical experience, not just the event itself but overhearing Kermit doing media interviews earlier in the day. He is real to me. Of course, he is real, as he is a real material object. He is also a real character – famous, popular and enduring. Yet neither objects nor frogs talk, run theatres, ride bicycles or do any of the other things that Kermit the Frog has done in his illustrious career. This is the power of the puppet: We know that it is not a living thing, yet we commit to it, believe in it and relate to it as if it was.

On a personal level, my motivation for wanting to witness Kermit's visit was simply that *The Muppet Show* is a fondly remembered part of my childhood and – as I colloquially phrase it – I have always 'had a thing' for Kermit. The opportunity to see him perform live (note, it is Kermit who is performing) was a childhood dream come true. On an academic level, I am fascinated by what it is that makes puppets so powerful, especially at a time when the digital is increasingly dominant. How can such a basic material object not only endure in its cultural significance but continue to find new forms and spaces of performance despite recent and growing trends to the technological and virtual? What makes a puppet so convincing or affectively powerful that an acknowledged object can make grown adults cry? How are the qualities of a puppet – its puppetness – employed to create new subjectivities and spaces,

DOI: 10.4324/9781003214861-1

and how might these be of academic as well as dramatic value? It is answers to these sorts of questions that I seek in a broad-ranging and long-running programme of research.

This volume presents the early stages of this research and is the first book-length geographical engagement with puppets. It builds upon a previously published paper to construct and consolidate a new area of human geographical endeavour: The geographies of puppets and puppetry. This introductory chapter is structured in three main sections, the first of which discusses the underrepresentation of puppets and puppetry in geography compared to other cultural forms and draws heavily on my previously published paper[1] (Banfield 2020). This section proposes both possible explanations for geography's under-engagement with puppets and reasons for urging fuller and deeper disciplinary engagement in the future. The second section introduces the grotesque and its relationship to puppets, articulates the distinctive nature of my engagement with puppets as grotesque and outlines the puppet spaces that are excavated and constructed in this volume. Finally, the third section details the overriding aims that have guided this research and presents an overview of the chapters that follow.

## Puppets and geography

Puppets are animated figures or objects, with common forms including glove or sleeve puppets, ventriloquists' dummies and marionettes. While there are overlaps between puppets and associated cultural forms from performing objects to automata, for the purposes of this book a puppet is distinct in being intended for performative use (so is not an object that is re-appropriated for such use) and requiring real-time manipulation by a human operator (so is not programmable for independent operation) (Francis 2012). While we might commonly associate puppets with children's cultural outputs, including television (*Sesame Street*), seaside holidays (*Punch and Judy*) or the theatre (*War Horse*, based on a children's novel), puppets are also used in popular adult cultural outputs, including their incorporation into televisual murder-mysteries (e.g., *Midsomer Murders*, *Endeavour*), fantasy fiction (e.g., *Rivers of London*) and horror films (e.g., *Robert*, *Puppet Master*). Puppets, then, are not confined to any one cultural medium or genre and are thus accessible from a range of analytical perspectives.

Despite this, the omission of puppets from academic engagement was acknowledged at least as long ago as the early 1990s, when their exemption from serious analysis was attributed to the commonness of puppets that rendered their analytic value effectively invisible (Schumann 1991). By the end of that decade, authors were commenting on a transformation in the cultural significance of puppets yet still lamenting scant scholarly interest with few puppet scholars and little engagement beyond puppet organisations (Kaplin 1999). Roll the clock forward another decade and although signs of diversification in those working academically with puppets was starting to be recognised, puppet literatures remained primarily directed towards specialist audiences (Francis 2012). At the start of yet another decade, puppets continue to fascinate audiences, find new forms and access new spaces of performance, yet academic engagement with them remains sparse, and geography exemplifies this well.

To explore disciplinary interest in puppetry compared to other cultural forms, I searched the academic database Scopus in August 2019, pairing key cultural search terms with 'geography': (1) within title, abstract and keywords and (2) as source title. These searches returned only 16 results for 'puppet + geography (title/abstract)' and only three for 'puppet + geography (source)', compared with 36 (title/abstract) and 12 (source) for 'cartoon', 48 (title/ abstract) and 22 (source) for 'comic', and 38 (title/abstract) and 7 (source) for 'filmic'. Given the long history of puppets compared to the more recent invention of printed and televisual media, with human use of puppets preceding theatre and organised religion (Francis 2012), this relative paucity of work on puppets is notable. This disciplinary neglect of puppets is highlighted further if we consider that even famous or celebrity puppets, such as Mr Punch and Pinocchio, do not rate a single mention but individual comic characters cropped up frequently in these searches, including Captain America, Superman and Tintin.

This is not to assert that there is no geographical work engaging with puppets, as there is no doubt consideration of Punch and Judy in the context of the emergence of the working class and its spatial association with the seaside, within historical geographies and/or coastal geographies. However, as puppets in such work do not feature in titles, abstracts and keywords, these more general engagements with puppetry are harder to trace (and are yet to be unearthed), making it difficult to establish a baseline for pre-existing disciplinary work. Even tourism geographies, which have addressed the changing face of live entertainment at the seaside seem not to acknowledge Punch and Judy as an archetypal live seaside entertainment (Gale 2005; Bull and Hayler 2009). By looking beyond geography, it is considerably easier to find work explicitly focusing on puppets with an emphasis on geographical themes, whether in terms of their cultural appropriation (Cutler Shershow 1994), their progressive re-spatialisation from the seaside to the shopping mall (Reeve and Reeve 2011) or their capacity to unpack the politics and performativity of the (disabled) body (Purcell-Gates and Fisher 2017).

This lack of disciplinary engagement is both unfortunate and surprising. It is unfortunate given growing public interest in puppetry despite the trend towards digitalisation and virtualisation of social and cultural life, suggesting an interesting tension ripe for geographical exploration. Specific puppet theatre shows have achieved iconic status (e.g., *War Horse*); super-sized characters periodically roam the streets to commemorate historic events (e.g., giant puppets in Liverpool to commemorate the Titanic) or to celebrate local identity (e.g., *Man Engine* in Cornwall); they continue to feature in new popular cultural outputs, from television (e.g., *Hold the Sunset*) to rap (e.g., 'Puppet' by *Tyler, the Creator*), and new puppetry festivals are being established (e.g., Nottingham Puppetry Festival in 2018).

It is surprising given the myriad ways in which puppetry intersects with disciplinary concerns and developmental trajectories, suggesting that geography has much both to contribute to and gain from engagement with puppetry. Puppetry is an internationally practised cultural form with a long history that is both culturally specific and globalising. For example, Bunraku (involving three puppeteers per puppet) originated in Japan but three-person ensemble puppetry is now a common feature of puppetry workshops in the UK, while Mr Punch was re-appropriated and renamed upon arrival in the UK from Italy. The sub-national level, too, exhibits a historically rooted geography of puppetry with certain locations being associated with specific types of puppetry (e.g., puppet opera in Sicily, and puppet film and animation in Prague), often uniting museums to preserve puppetry artefacts and contemporary facilities to sustain the practice into the future. As a socially, culturally, economically and politically situated practice, puppetry is inevitably implicated in socio-political processes, tensions and transformations from the community to the global geopolitical level. Recognising this situatedness, my own research focuses on contemporary Anglo-American cultural uses of puppets to establish a baseline of understanding within my own cultural milieu.

Puppetry is, furthermore, an inherently spatial practice, occurring in performance contexts ranging from theatre, film and television productions to community events within street, carnival and festival settings. These spaces are transformed by puppet performances, prompting spectators to question and experience them anew as, for example, War Horse exits the theatre and canters down the street, or a human-scaled cityscape is dwarfed by the presence of a giant marionette, making puppets immediately relevant to non-representational concerns in generating powerful affective responses. Simultaneously, socialities are changed as the puppet rather than the puppeteer becomes the focus of attention, fuelling the agency, subjectivity and sociality of the puppet, while also drawing attention to the conjoined puppet–puppeteer bodyscape as its own space, which exhibits a different geography according to whether the puppet is operated directly (in a glove) or remotely, whether from below (using rods) or above (using strings) and whether the puppeteer is intended to be a powerful visible presence within the bodyscape or eclipsed by the puppet. Moreover, this embodied relation between puppet and puppeteer and the ability to transgress bodily norms with or violate the body of the puppet, lend themselves to contemporary interests in embodiment, materiality and more-than-human relationality. Not only do the varied configurations of puppet–puppeteer bodyscape provide new perspectives on more-than-human entanglements, but the relationship between puppetry and masked performance invites new work on the significance or otherwise of faciality in relating and communicating both within human and more-than-human socialities. Similarly, the diverse material forms that puppets can take and their different affordances and limitations offer alternative ways to explore the potential for non-linguistic communication, including mime and gesture, especially in relation to differently abled bodies, while questions surrounding what it is that makes a puppet performance convincing and how we sustain or fail to sustain the believability of the puppet, are also beguiling for non-representational geography. As such, puppetry nestles in the intersections between the visual, the verbal, the performative, the material, the more-than-human, the emergent and the affective, and is amenable to investigation from cultural, socio-historical, political and non-representational perspectives.

While surprising, this lack of geographical engagement with puppets and puppetry is not altogether beyond explanation, as at least three factors might help to account for puppets failing to find favour within the discipline. The first of these is the puppet's status as a 'low' form of culture, being a simplistic, live entertainment associated with the crafts rather than the arts and with the working class (Leach 1983), showing strong connections to folk and popular theatre (Bell 1999) and being commonly perceived as children's entertainment (Gocer 1999/2000; Crone 2006). This characterisation might in part account for the relative lack of geographical interest in puppets given the discipline's historic endeavour to establish itself not only as a serious academic discipline but also as a formal and 'hard' science (Gregory 1978). Second, the puppet's deep historic roots predate the rise of Enlightenment thinking, at which time puppets could be considered a degenerate form of a god (Jurkowski 2013), potentially contributing to their exclusion from a discipline that sought for decades to distance itself from such supposedly irrational, non-Western, traditional (read, archaic) practices. Third, puppets have also traditionally been associated with marginalised sections of society, often travelling or itinerant communities, with troupes of puppeteers moving from town to town on a periodic and temporary basis (Cutler Shershow 1994; Churchill 2014), further mitigating against their inclusion in formal scholarship.

Disciplinary developments, however, show that such a view is outdated, and contemporary sensitivities encourage a re-evaluation of puppets as a legitimate and enticing geographical concern. The humanistic turn and the rise of new cultural geography means that positivist scientific approaches are no longer considered the only route to knowledge, and with the rise in interest in performance and practical forms of knowledge in non-representational geography has come growing appreciation of all kinds of practices, opening the door to engagement with the arts and not just those associated with elite sections of society (Anderson and Harrison 2010). Meanwhile, the broadening of interests and knowledges considered valid within geography means that children's geographies are now a rich, established and evolving field of inquiry (Holloway 2014), so even if it was appropriate to ascribe puppets only to children, that would no longer warrant their exclusion from geographical endeavour. Similarly, geography's surging interest in the practices, politics and experiences of diverse mobilities (Sheller and Urry 2006) makes the current moment timely for a renewed appreciation of mobile communities and lifestyles, such as those historically associated with puppetry, while ongoing efforts to decolonise geography and democratise Western academia mean that alternative forms of knowledge are increasingly legitimised as meaningful practices of world-making (Radcliffe 2017).

All these disciplinary developments encourage the validation of puppets and puppetry as not only legitimate but fruitful and long overdue avenues for geographical inquiry. Puppets are both a representational and practical form of knowledge, encompassing aspects of both art and craft, thus making them amenable to inquiry from a range of epistemological perspectives and analytical approaches. Moreover, as a feature of human society – both child and adult, both marginal and mainstream – around the world and throughout history, they simultaneously serve as conduits of cultural consistency and axes of cultural accentuation, establishing both equivalence of validity and uniqueness of form and style between cultural contexts. Puppets, then, are well-suited to contemporary disciplinary interests and this volume sets the scene for broader geographical engagement with puppets, by focusing on the representations and uses of puppets in diverse cultural forms, through which initial understandings, analytical frameworks and conceptualisations of the cultural geographies of puppets are generated.

## Puppets and the grotesque

The ontological paradox of the puppet is that it is simultaneously a material object and a signifier of life (Tillis 1996), suspended between being animate and inanimate, subject and object (Zamir 2010; Cappelletto 2011). The spectator can sustain, lose or fluctuate between the belief in the illusory vitality and awareness of the material reality. While different approaches might be adopted by those staging the puppet performance, either masking or making clear the artifice behind the illusion (Francis 2012; Jurkowski 2013), the illusion is never perfect (Williams 1991) so the spectator commonly alternates between belief and awareness, illusion and reality. Lacking a formal scientific label, this phenomenon goes by several names, including oscillation, opalisation and double-vision (Williams 1991; Gross 2011; Francis 2012; Jurkowski 2013). Full of contradictions, puppets make us believe and not believe at the same time (Bicât 2007), but those very contradictions make it difficult to sustain both the belief and the awareness, hence the oscillation.

As my interest in this book lies primarily in representations of puppets, my analytical focus is on the puppet/human distinction and the bodily border between them rather than attending to the illusion–reality or subject–object contradiction, (although these both feature). This does include the distinction and relation between the puppet and the puppeteer (e.g., in Part 2) but it also includes those between a fictionalised character and their puppetised plight at the mercy of another (in Part 1) and the transformations that are possible between puppet and human in cultural works (in Part 3). Thus, while it might be true that the most important factor in any theory of puppet theatre is the relationship between the puppet and the puppeteer (Jurkowski 2013), my attention is directed to representations and uses of puppets in cultural contexts beyond the theatre and – as such – attends to human–puppet relationships beyond that of puppet and puppeteer. This broader perspective reflects the fact that I write not as a professional puppeteer or puppet maker or theatre maker, but as a puppet enthusiast who as an academic geographer has identified interests common to both puppetry and geography. Two such interests are the bodily relation between humans and puppets (in the varied forms above) and the reshaping of the world around the border between them (in various configurations of border crossings). Thus, I take seriously the suggestion that what is interesting is what is discovered at the intersection between the material puppet and the human body (Mello 2016) and I adopt an explicitly geographical attitude towards that intersection in considering that bodily border as a key site for the making of new and uniquely puppet worlds.

One concept that relates directly and strongly to combinatory forms such as human–object hybrids is the concept of the grotesque, and while this is not a book about the grotesque, I do draw heavily on this idea while also applying and developing the idea of the grotesque in novel contexts. Notable for its lack of definitional clarity (Kayser 1963; Barasch 1971; Duggan 2016), and a proposed dearth of scholarship explicitly concerning the grotesque (Connelly 2012), the grotesque is – nonetheless – a useful and productive concept. It is characterised by a multiplicity of uses and applications, which have evolved over time and been modified in culturally specific contexts (Kayser 1963; Barasch 1971; Harpham 1976, 2006; Li 2009; Connelly 2012; Duggan 2016). Most commonly considered an aesthetic term or mode, its history is easier to trace than its practical or phenomenological occurrence, with the origins of the term being rooted in the excavation of wall ornamentation in late-fifteenth-century Italy but with the existence and practice of the grotesque stretching much further back into history (Harpham 1976; Li 2009; Pilný 2016).

As an aesthetic means of questioning received norms and standards of cultural expression, it has been variously associated with both artistic freedom and deviancy (Harpham 1976; Li 2009; Connelly 2012; Duggan 2016), and stereotypical expressions of the grotesque involve monstrous forms of human–animal–plant hybrids often involving grossly exaggerated features and a lack of order or organisation (Kayser 1963; Harpham 2006; Duggan 2016). The potential of the grotesque to subvert social norms and hierarchies has been much discussed (Danow 1995; Li 2009; Duggan 2016), with its emancipatory potential often being grounded in the relationship between the grotesque and the carnival, which – rooted in Bakhtin's (1984) work – sees the carnival as an alternative space-time in which everyday social relations, hierarchies and norms are suspended and new social forms can emerge. However, such affirmative views have been critiqued for romanticising the carnival and its subversive potential (Danow 1995; Li 2009; Duggan 2016), and overlooking the psychologically repressive and alienating power of the grotesque (Kayser 1963; Russo 1994; Goodwin 2009) as something fearful rather than playful, which brings us to the key feature of the grotesque: Contradictions.

This dichotomy of views highlights a division within the grotesque between permutations that prioritise the comic or the terrible (Barasch 1971; Harpham 1976; Danow 1995; Pilný 2016), and it is the intersection of the two – as in tragicomedy – that perhaps most clearly evokes the grotesque (Harpham 1976). The grotesque speaks to an ambiguous mix of hilarity and terror and a tension between attraction and repulsion (Harpham 1976, 2006). Whether seen as an aesthetic mode, a source of political potential or an unsettling affective experience, the grotesque relies upon contradictions, whether between incompatible conceptual categories, norms and the abnormal or contrasting affective responses. These contrastive structures are both a characteristic and a theme of the grotesque, which generates boundary creatures forged through the fusion of incompatible elements (Kayser 1963; Harpham 1976; Danow 1995; Connelly 2012). Moreover, it is the utilisation of such fusions to question what it is to be human through the juxtaposition of different categories or realities that epitomises the grotesque (Connelly 2012; Pilný 2016). The grotesque, then, is more a verb than a noun (an action or process), which is always unfinished but capable of critiquing and mocking society and of expressing and revealing profound truths by rendering conventional categories inapplicable and generating new ways of ordering and knowing the world (Kayser 1963; Harpham 2006; Connelly 2012; Pilný 2016).

The qualities of the grotesque have been articulated as entailing excess, exaggeration, degradation and changes of scale (Bakhtin 1984; Li 2009; Duggan 2016) or as involving aberration, combination and metamorphosis (Connelly 2012). These two sets of qualities are not mutually exclusive, however, as combination might involve exaggeration and metamorphosis might entail degradation, but common to all these approaches to the grotesque is a focus on the body, especially the human body and its processes and excesses (Duggan 2016). The constellation of grotesque concerns for the body, excess, disgust, social norms and so on, has led to an academic debate as to the sexist nature of the body in the grotesque, with some authors highlighting the framing of the female body as the embodiment of all that is grotesque and a source for the grotesque-ing of male bodies through the incorporation of female bodily attributes (Russo 1994; Connelly 2012), and others critiquing such perspectives as too singular and universalising (Li 2009; Duggan 2016).

Aside from this feminist debate, several facets of the body in the grotesque are worth dwelling on briefly here. The first thing to note is that the body in the grotesque involves both bodily and psychological excesses (e.g., oversized facial features or inebriation) (Duggan 2016) and that physical deformity of the human body has been linked in the grotesque to spiritual and/or intellectual deviancy (Harpham 1976). Moreover, this emphasis on physical deformity is identified as one of two dominant treatments of the body in the grotesque, the other being the disavowal of human agency in favour of an external mechanical force mobilising the body (Harpham 1976). Bakhtin's work again becomes central here, as it draws out at least four key aspects of the body in the grotesque: Firstly, the emphasis is on the fusion of body and world and a questioning of the limits of the human body and secondly, there is an abiding concern for the boundary between bodies, constituting a two-body image through the fusion of a body and a material form (a body grotesque). Thirdly, the human body is topographical in both form and movement, drawing moral as well as anatomical distinctions between the upper and lower body and ascending and descending movements and fourthly, emphasis is placed on specific anatomical features or facilities, such as that which protrudes from the body (e.g., arms, hands), that which outgrows itself (the bowels, phallus) and that which leads beyond or reaches within (e.g., nose, mouth) (Bakhtin 1984). In stark contrast to the singular, clearly bound classical body, the grotesque body is open, social and unfinished, excessive and breaking free from conventional regulation (Russo 1994; Li 2009; Duggan 2016).

Two key aspects of the above summary are pertinent for this volume's focus on puppets: The combination of human and non-human forms and the preponderance of treatments of the human body that render human agency mechanical or externally manipulated. However, puppets are of relevance to the grotesque for other reasons, which, in turn, will lead to an articulation of how I am employing the grotesque in specific ways. For one thing, common associations have been drawn between carnival, the *commedia dell'arte* and puppets in grotesquery (Kayser 1963) and puppets have been explicitly identified as an enduring theme within the grotesque (Kayser 1963; Harpham 1976; Pilný 2016). It has also been noted that important instantiations of the grotesque in modern theatre have been rooted in puppets, for example through the presentation of the world as a puppet universe, portraying humans as puppets, or depicting life as meaningless as if it were a puppet play (Pilný 2016). Furthermore, the combination of tragedy and comedy, so characteristic of the grotesque, is also a definitional feature of possibly the most famous puppet of all – Mr Punch – who himself often forms the aforementioned nexus between carnival, *commedia dell'arte* and puppets.

Puppets, then, are perfectly at home in the grotesque, and the puppet–puppeteer configuration is a two-body image in the vein of Bakhtin (1984), constituted through the fusion of the human body and the material form. In turn, the grotesque provides fertile conceptual terrain for the analysis that follows, but my own engagement with the grotesque is distinctive in some important ways. Firstly, I am applying ideas and understandings of the grotesque to a specific example of the grotesque (the puppet) in its varied constructions, and I am interrogating not what it is to be human but what puppetness is and what it facilitates. In this context, I see puppetness not as the homogenised formal features of puppetry (Francis 2012) but as a target for interrogation in the generation of particular/peculiar subjectivities and spatialities that is rooted in the human/puppet border at the heart of the puppet's ambiguous status and grotesque nature. Moreover, I am not considering the grotesque as a space but the puppet as a space, which is variously constructed through different instantiations of the grotesque in the diverse uses and configurations of 'puppet' that are explored. I am also concerned not solely or even primarily with the puppet–puppeteer relationship, but with multiple ways in which humans and puppets relate in popular cultural constructions, whether that be humans being presented as metaphorical puppets, puppets being possessed by the spirits of humans, human–puppet transformations, or anything else. Thus, I draw on the grotesque in specific ways, with the analytical and spatial focus on the puppet, to craft a cartography of the puppet as constructed in its diverse grotesquery. I attend explicitly to the blurring of the human–puppet border, conceptualised as a practice of borderscaping, in the construction of the puppet through which various gaps between contradictory aspects of human and puppet are bridged, accounting for the diverse array of puppet constructions and spatialities.

## Aims, approach and chapter summary

Through a geographical analysis of the uses of puppets in a range of popular cultural forms (literature, television, film, music and theatre), I aim to identify and conceptualise different aspects of puppet constructions both in their specificity within particular cultural forms and in their totality across cultural forms. Thus, while I begin with an everyday fictional construction of the puppet, this is swiftly set in the context of a contradictory practitioner perspective, and each chapter further pluralises and complicates this initial understanding. I also aim to situate and understand these aspects and conceptualisations in relation to each other, so although the individual chapters are discrete and some of their findings are quite divergent, they are increasingly brought into alignment at the conceptual level, through the development of overarching frameworks that can accommodate considerable diversity of puppet constructions in an organised fashion.

My starting point for this conceptual work is the border, or more specifically borders, between humans and puppets that are muddled in the construction of puppets as an integrated puppetised form. Part 1 identifies a range of such borders, which are developed in Part 2 into the idea of the borderscape, drawing on literature from border studies, before then applying and extending that conceptualisation in Part 3 in the context of puppet–human and human–puppet becomings, or transformations. Borders are a distinctly spatial phenomenon and border studies is a field of scholarship experiencing considerable growth and diversification, developing from a sub-discipline of political geography into an interdisciplinary field and adopting more dynamic conceptualisations of borders as evolving, socially constructed spaces of political potential and sites for investigation and reimagination (Wastl-Walter 2009; Brambilla 2015; Krichker 2020). It is in this latter spirit that I engage with borders and borderscapes, albeit in a very different disciplinary context (cultural geography) compared to more conventional social and political arenas (Brambilla 2015), and favouring the term borderscape over borderlands (Wastl-Walter 2009) or border zones (Krichker 2020). Rather than attending to people who regularly cross or live in the border (Krichker 2020), I focus on how borders of a different sort – human/puppet

borders – are used in the construction of puppets and the spatialities generated through these constructive processes of becoming.

Becoming is another term of specific disciplinary interest and is here taken
to be a transformational process generative of entities, spaces and subjectivities (that we would conventionally assume to be pre-existing) as emergent
assemblages across materialities, mobilities and corporealities, both perceptual and affective (Wylie 2002; McCormack 2009a). This becoming-together
unsettles boundaries between seemingly discrete entities (McCormack
2009a), reminiscent of the muddling of borders in puppet constructions.
With posthuman (and grotesque) attentiveness to the boundary of the human
through reading the figure of the puppet, I inquire as to the emergence,
capacities and spatialities of the puppet in its construction as an entity (as
state) in Part 1, as a process of spacing (borderscaping) in Part 2 and as
transformational (as becoming) in Part 3.

Spatiality – or space – is understood here not as a static container or background for action but as an anti-essentialist and active process of spacing,
giving rise to 'spaces' (as we conventionally conceive them) as provisional
outcomes of these processes of becoming among myriad human–nonhuman
relations, rather than fixed and finished. Thus, puppet and human are inter-
spliced across the puppet/human bodily border in the constructive becoming
of puppets, which is simultaneously the becoming of novel puppet spatialities (Part 1). This process of intersplicing is spatialised through the notion of
the borderscape (Part 2), while these puppets and their spatialities are further
reconfigured when puppets become humans and humans become puppets
(Part 3). This, then, is the overall structure and direction of this work.

As a geographer, I seek not only to describe the diverse and peculiar spatialities that are afforded by these borderscaping practices and explore their
implications (e.g., for the integrity of the character or tale and for novel
approaches to literary criticism in geography) but also to do similar with the
grotesque, thereby extending and refining conventional conceptualisations of
the grotesque from an explicitly geographical perspective, as well as contributing to geography itself. However, despite this disciplinary perspective and
the forging of analytical links with diverse disciplinary interests (e.g., geocriticism, scale, affect, social critique, corporeality, posthumanism), these
links are intentionally foregrounded in the conclusion to each part, rather
than being fully elaborated in individual chapters, so that the emphasis in the
chapters can remain focused on puppets, the grotesque, how they are each
constructed and the spatialities that emerge. That said, the disciplinary
engagement develops as the work progresses, incorporating consideration of
– for example – the intersection of biopolitics and geopolitics at the level of
the individual, comic book geographies and musical geographies, non-representational geography and affect, critical geography and the impact agenda,
and more-than-human geography, thereby establishing multiple and diverse
contributions.

The book consists of three parts, delivering a progressive unearthing of the
borders (Part 1), borderscapings (Part 2) and becomings (Part 3) that are

found to constitute the unique and strange – peculiar – geographies of puppets. At the same time, each part adopts a different perspective on puppets, beginning with metaphorical references to puppets in Part 1, through performative uses of puppets in Part 2, to the transformative potential of puppets in Part 3. Along the way, the understandings of puppets that are generated become progressively more complex and more deeply and broadly embedded in the discipline of geography, progressing from a quotidian understanding with relevance to networked ideas of space in Part 1, through appreciation for the affective power of puppets in Part 2, to concern for the relevance of ideas of multicorporeality to current debates pertaining to posthuman subjectivity in Part 3.

Part 1 – *Metaphorical Puppets* – addresses the use of puppets in adult fiction as metaphors for human life and society and is concerned with the question 'what is a puppet?', not in the sense of dictionary definitions or types of puppet but in the sense of an everyday culturally assumed understanding of 'a puppet'. Chapter 1 focuses on passing metaphorical references to establish a quotidian and composite understanding of 'a puppet' that encompasses a specific bodily form (a ventriloquist dummy's head on a marionette's body) and mode of control (the strings of a marionette), establishing that this version of the grotesque incorporates three elements (one human and two distinct types of puppet). Consistent with the grotesque, core contradictions emerge as central to this metaphorical use of puppets, relating to both the subjectivity and materiality of the puppet, with implications for the reader's capacity to empathise with the plight of the puppetised character and generative of four distinct puppet spatialities.

Chapter 2 attends to narrative uses of puppets as a compositional device, which emphasise psycho-social over bodily forms of control and explore in more depth this relationship of control. Three aspects of this analysis are considered in explicitly spatial terms: The strategies by which control is established, the attitudes or orientations that a puppet master takes towards its victim and mechanisms of puppet-to-puppet-master integration, which are used to suggest that rendering congruous can be as important to the grotesque as the combination of incongruous elements. Together, these chapters establish a quotidian understanding of the puppet and their literary use as social commentary, highlight a number of borders at the crux of their narrative employment and draw out a range of relational spatialities and spatialising processes that emerge from such uses of puppets. Moreover, it proposes the 'character contorted' as a supplementary term for the body grotesque, to accommodate more explicitly psychological constructions of puppets as grotesque. The broad disciplinary themes to which this part speaks, and which are elaborated in the Conclusion to Part 1, include scaling effects in relation to the space of the body within a networked sociality, the critical potential of puppets and the individualisation of geopolitics and biopolitics within interpersonal rather than societal contexts.

Part 2 – *Performative Puppets* – addresses the substantive use of puppets as performers or characters in a range of cultural forms (literature, film, theatre and music) and poses the question 'what grants puppets their performativity?' to tease out how puppets 'work'. Performativity here is taken to be the quality of a practice as generative rather than reflective of the world (McCormack 2009b; Francis 2012), contributing to its continual becoming rather than maintaining its existing being. Chapter 3 addresses such uses of puppets in literary and filmic outputs to establish a dimensional framework of puppet use and identify the interrogation of the human–puppet border as a key site for the generation of these puppet spatialities. In this context, the grotesque is considered not to lie in the gaps between conceptual categories but in the very bridging of them at the border which connects as much as it divides, by reconfiguring the spatial understanding of the gap (i.e., difference-as-distance) into a topological gulf that is bridged by the very relations of difference that hold conceptual categories apart yet connected. The peculiarity of the spatialities generated through such bridging is related to the degree to which puppetness is explicitly interrogated in the telling of the story, which is subsequently conceptualised as a process of borderscaping through the intercorporation of puppet and human as an essential yet mutually incompatible pairing of eliminating puppet/human distinctiveness and foregrounding their distinctiveness.

Chapter 4 attends to puppets in musical contexts, including their associations with musical theatre and musicians and their representations in musical works. Through an analysis of music and lyrics from opera to rap, two overarching themes emerge: Mood and movement, and tensions and contrasts. The construction of puppets through evocations of mood and movement is articulated as twin perceptions of grounded lightness generated through exaggerated jauntiness and anatomical incoherence through either disassembly or solidification of an otherwise multi-limbed anatomy. Additional topological gulfs are unearthed, and an especially creative combination of body grotesque and character contorted is identified in an operatic redistribution of Mr Punch's subjectivity to the percussive background rather than his personal lyrical and musical articulations, thereby simultaneously hyper-puppetising Mr Punch as an archetype and de-individualising him as a character.

Chapter 5 focuses on the use of puppets for specific dramatic and affective purposes. The chapter begins by teasing out the significance of specific puppet types for their performativity by comparing two styles of puppet that are similar in visual appearance (brightly coloured fur and ping-pong eyes) but that use the puppet–puppeteer relationship very differently (with the puppeteer concealed in one but obvious in the other). This uncovers an important pair of topological gulfs between faciality and pedality on the one hand and excessiveness and absence on the other hand, with different combinations of emphasis constituting different ways of rendering puppet and human equivalent. Subsequently, the potential and power of puppets for moral messaging and affective impact is explored through a detailed comparative analysis of the same story in different cultural media (e.g., *War Horse* as novel, film and puppet stage-play). A detailed discussion of divergent approaches to the borderscape draws out further developments in our conceptualisation of how puppetness is used: Radical dissolution, theatrical dissolution, theatrical strengthening and radical strengthening of the puppet–puppeteer border, which are understood as strategies of borderscaping. Further, a distinction (topological gulf) is drawn between the border and the borderscape wherein the integrity of the puppet–puppeteer borderscape can be maintained even as the border is variously maintained or dismantled, and this is related to the uncanniness of the puppet: The fluctuation between believing and disbelieving the vitality of the puppet. Bringing the border-borderscape relationship into conversation with this uncanniness draws attention to the construction of the puppet on both explicit (cognitive) and implicit (pre-reflective) registers, with these spatialised at the border and across the borderscape, respectively. It is this latter disciplinary (non-representational) context and the centrality of affect to the construction and reception of puppets that form the focus for the elaboration of the geographical relevance of Part 2 in its conclusion, which also presents a revised dimensional framework that incorporates the supplementary findings from Chapters 4 and 5.

Part 3 – *Transformational Puppets* – addresses the ontological possibilities of human–puppet transformations in the context of disciplinary interest in becoming and is guided by the question: 'How are puppets constructed?' to develop a more detailed and integrated understanding of these constructions than that which has emerged so far. Chapter 6 focuses on the transformation from puppets to humans as represented in multiple illustrated tellings of Pinocchio in relation to his materiality, sociality and metamorphoses. This visual analysis identifies a series of constructive contradictions on which Pinocchio is founded, which – in turn – are associated with multiple micro-spatialities that emerge from the borderscaping practices in each work. Although these micro-spatialities remain implicit in the narrative, they can – if inconsistent – detrimentally affect both the construction of Pinocchio and the integrity of the tale, such that the borderscape underpinning a puppet construction can be self-limiting. This chapter both develops our understanding of puppetness and schematises these contradictory constructions, while also suggesting an alternative spatial means of analysing literary works.

Chapter 7 attends to the transformation from humans to puppets across multiple works. It confirms the broad applicability of the emergent understandings of puppetness from the preceding chapters while also developing and refining those understandings further. In addition to the contradictions characteristic of puppet constructions, a series of constructive complementarities are identified in these constructions, suggesting an alternative way of thinking about the grotesque. Ten distinct ways are identified in which the human transformation into puppet is crafted, which are organised into five paired axes of progressive transformation. In turn, this sets the notion of intercorporeality across the human–puppet bodily border within a broader context, bringing into consideration issues of affinity, solidarity and intertextuality. Conceptualising these developments in spatial terms, intercorporeality is found to sit alongside epicorporeal (non-bodily but individual) devices and within a broader pericorporeal (relational or contextual) setting, which can be employed in isolation or in combination, with the latter proposed in its maximum form as constituting a pancorporeal ontological space of possibility. This is subsequently rearticulated as a series of interwoven technologies (more associated with puppets) and ecologies (more associated with humans) within this multicorporeal framework, which – in turn – is used to flesh out and formalise the notion of the borderscape and the practice of borderscaping.

The Conclusion to Part 3 considers the applicability of the findings from Chapters 6 and 7 to the other chapter to present a composite diagram of the ecologies and technologies of puppet-becomings as a constructional framework to supplement the dimensional framework from Part 2, and relates this constructional framework to both the grotesque and the borderscape. Finally, two core synergies between this integrative work and disciplinary interests are unpacked, in relation to geocriticism and a posthumanist subjectivity. An alternative understanding and practice of geocriticism – exploring the real/ imaginary spaces and spatial practices of literature (Tally Jnr 2011; Westphal 2011) – is outlined, which attends to the spatialities implicit rather than explicit in a narrative, and the discussion of subjectivity in Chapter 7 is extended to consider the specific benefits of puppets to geographical concerns about the need for – and the perils of – decentring the human in these current posthuman times, reinforcing and further specifying the particular potential within puppets and their unique peculiarities for geographical scholarship.

Finally, the conclusion provides answers to the three guiding research questions as a recap on the preceding content, and articulates empirical, methodological, analytical and conceptual contributions. Beyond the initial emphasis on puppets, I narrate the implications of these chapters for understandings of the grotesque, before turning my attention more explicitly to geography. The myriad puppet spaces and spatialities that emerge through the chapters are considered (organised around five sets of spatiality from the body to the border) and the broader disciplinary contributions are fleshed out further, including the implications of this research for the posthuman subject, a diversification of non-representational theoretical perspectives and advocating the development of the geographies of both puppets and the grotesque. I bring this first book-length geographical foray into puppets to a close by reaffirming my conviction that puppets are not only quintessentially spatial but also of considerable and growing disciplinary relevance, with the potential to advance geographical scholarship and impact in multiple directions. Leaving puppets as a disciplinary blind spot is simply not an option. To turn Mr Punch's famous phrase on its head, that would very much *not* be the way to do it.

## Conclusion

This introduction has established that puppets continue to be significantly underappreciated in geography despite their relevance to multiple and diverse streams of contemporary geographical scholarship. This volume seeks to redress that neglect through a sustained exploration of how puppets are constructed in popular culture. Although the starting point for my exploration is the representation of puppets, my emphasis on how puppets are constructed draws out the active, practical aspect of representation, and these representational constructions are intimately interwoven with other non-representational considerations such as powerful affective responses, the rendering of puppets through mood and movement and the puppet as uncanny. Thus, alongside the examination of how puppets are constructed, this work highlights the non-representational nature of representation.

The analytical emphasis progresses from thinking about the different border-crossings between human and puppet in the construction of puppets as entities, through an interrogation of these border-crossings as a spatialised practice of borderscaping, to an exploration of the construction of puppets as transformational entities in their process of becoming. Along the way, I establish an array of conceptualisations of the construction of puppets, which are – in Part 3 – integrated into a holistic overview, while also generating conceptual developments with respect to the grotesque, answers to the guiding research questions and considerable contributions for geography (and beyond). Ultimately, puppets are confirmed as inherently spatial, being constituted in specific spaces, constituting their own unique spatialities and re-spatialising the grotesque. Puppets are affirmed as valid and valuable members of the geography community and – in their co-implication with the grotesque, the borderscape and the uncanny – provide a rich resource for disciplinary excavation in forging grotesque geographies of diverse borderscapes.

## Note

1 Part of this introduction – the section on puppets and geography – originally formed part of the introduction to a paper first published in *Cultural Geographies* (see Banfield 2020). That material has been extended slightly for inclusion in this volume, and the referencing style has been modified to ensure consistency throughout this work.

## References

Anderson, B. & Harrison, P. (2010) The promise of non-representational theories. In: B. Anderson & P. Harrison (eds) *Taking place: Non-representational theories and geography*. Ashgate Publishing Ltd: Farnham and Burlington, 1–34.

Bakhtin, M. (1984) *Rabelais and his world*. Indiana University Press: Bloomington, IN.

Banfield, J. (2020) 'That's the way to do it!': Establishing the peculiar geographies of puppetry. *Cultural Geographies*, 28 (1): 141–156. doi:10.1177/1474474020956255.

Barasch, F.K. (1971) *The grotesque: A study in meanings*. De Gruyter Mouton: The Hague.

Bell, J. (1999) Introduction: Puppets, masks and performing objects at the end of the century. *The Drama Review*, 43(3): 15–27.

Bicât, T. (2007) *Puppets and performing objects: A practical guide*. The Crowood Press: Marlborough.

Brambilla, C. (2015) Exploring the political potential of the borderscapes concept. *Geopolitics*, 20(1): 14–34.

Bull, C. & Hayler, S. (2009) The changing role of live entertainment at English seaside resorts at the beginning of the twenty-first century. *Tourism Geographies*, 11: 281–307.

Cappelletto, C. (2011) The puppet's paradox: An organic prosthesis. *RES: Anthropology and Aesthetics*, 59/60: 325–336.

Churchill, D. (2014) Living in a leisure town: Residential reactions to the growth of popular tourism in Southend 1870–1890. *Urban History*, 41(1): 42–61.

Connelly, F.S. (2012) *The grotesque in Western art and culture: The image at play.* Cambridge University Press: New York.

Crone, R. (2006) Mr and Mrs Punch in nineteenth-century England. *The Historical Journal*, 49(4): 1055–1082.

Cutler Shershow, S. (1994) "Punch and Judy" and cultural appropriation. *Cultural Studies*, 8: 527–555.

Danow, D. (1995) *The spirit of carnival: Magical realism and the grotesque.* The University Press of Kentucky: Lexington, KY.

Duggan, R. (2016) *The grotesque in contemporary British fiction.* Manchester University Press: Manchester.

Francis, P. (2012) *Puppetry: A reader in theatre practice.* Palgrave Macmillan: Basingstoke.

Gale, T. (2005) Modernism, post-modernism and the decline of British seaside resorts as long holiday destinations: A case study of Rhyl, N Wales. *Tourism Geographies*, 7: 86–112.

Gocer, A. (1999/2000) The puppet theatre in Plato's Parable of the Cave. *Classical Journal*, 95(2): 119–129.

Goodwin, J. (2009) *Modern American grotesque: Literature and photography.* The Ohio State University Press: Columbus, OH.

Gregory, D. (1978) *Ideology, science and human geography.* Hutchinson: London.

Gross, K. (2011) *Puppet: An essay on uncanny life.* University of Chicago Press: Chicago, IL.

Harpham, G. (1976) The grotesque: First principles. *The Journal of Aesthetics and Art Criticism*, 34(4): 461–468.

Harpham, G. (2006) *On the grotesque: Strategies of contradiction in art and literature.* The Davies Group Publishers: Aurora, CO.

Holloway, S.L. (2014) Changing children's geographies. *Children's Geographies*, 12(4): 377–392.

Jurkowski, H. (2013) *Aspects of puppet theatre*, 2nd edition. Palgrave Macmillan: Basingstoke.

Kaplin, S. (1999) A puppet tree: A model for the field of puppet theatre. *The Drama Review*, 43(3): 28–35.

Kayser, W. (1963) *The grotesque in art and literature.* Indiana University Press: Bloomington, IN.

Krichker, D. (2020) They carry the border on their backs: Atypical commerce and bodies' policing in Barrio Chino, Melilla. *Area*, 52(1): 196–203.

Leach, R. (1983) Punch and Judy and the oral tradition. *Folklore*, 94(1): 75–85.

Li, M.O. (2009) *Ambiguous bodies: Reading the grotesque in Japanese Setsuwa Tales.* Stanford University Press: Stanford, CA.

McCormack, D.P. (2009a) Becoming. In: R. Kitchin & N. Thrift (eds) *International encyclopedia of human geography.* Elsevier (online resource), 277–281. Accessed 11 July 2021.

McCormack, D.P. (2009b) Performativity. In: R. Kitchin & N. Thrift (eds) *International encyclopedia of human geography*. Elsevier (online resource), 133–136. Accessed 11 July 2021.

Mello, A. (2016) Transembodiment: Embodied practice in puppet and material performance. *Performance Research*, 21(5): 49–58.

Pilný, O. (2016) *The grotesque in contemporary Anglophone drama*. Palgrave Macmillan: London.

Purcell-Gates, L. & Fisher, E. (2017) Puppetry as reinforcement or rupture of cultural perceptions of the disabled body. *Research in Drama Education: The Journal of Applied Theatre and Performance*, 22: 363–372.

Radcliffe, S.A. (2017) Decolonising geographical knowledges. *Transactions of the Institute of British Geographers*, 42(3): 329–333.

Reeve, A. & Reeve, M. (2011) Punch and Judy at the beach and in the mall. *Visual Culture in Britain*, 12: 17–31.

Russo, M. (1994) *The female grotesque: Risk, excess and modernity*. Routledge: New York.

Schumann, P. (1991) The radicality of the puppet theatre. *The Drama Review*, 35(4): 75–83.

Sheller, M. & Urry, J. (2006) The new mobilities paradigm. *Environment and Planning A*, 38(2): 207–226.

Tally, R.T. Jnr (2011) Introduction: On geocriticism. In: R.T. Tally Jnr (ed) *Geocritical explorations: Space, place and mapping in literary and cultural studies*. Palgrave Macmillan: New York, 1–9.

Tillis, S. (1996) The actor occluded: Puppet theatre and acting theory. *Theatre Topics*, 6(2): 109–119.

Wastl-Walter, D. (2009) Borderlands. In: R. Kitchin & N. Thrift (eds) *International encyclopedia of human geography*. Elsevier (online resource), 332–339. Accessed 11 July 2021.

Westphal, B. (2011) *Geocriticism: Real and fictional spaces*. Palgrave MacMillan: New York.

Williams, M. (1991) Aspects of puppet theatre/the language of the puppet. *Australasian Drama Studies*, October 1, 19: 67–75.

Wylie, J. (2002) Becoming-icy: Scott and Amundsen's South Polar voyages, 1910–1913. *Cultural Geographies*, 9: 249–265.

Zamir, T. (2010) Puppets. *Critical Inquiry*, 36 (3): 386–409.

# Part 1
# Metaphorical puppets

# Introduction to Part 1

The question guiding this first part is – at first glance – a simple one: What is a puppet? However, it will swiftly become clear that this question is not so simple, after all. Indeed, it will not be answered comprehensively in this first part, which provides only a starting point for a much fuller response that will be formulated through the rest of the volume. My intention in this first part is to establish an everyday, naïve understanding of what a puppet is, which will subsequently be expanded and deepened. This understanding will be everyday in the sense that it is commonly held and accepted, such that if someone uses the term puppet to describe somebody or something, the meaning of such a description is comprehensible and assumed to be comprehended in the manner intended. It is naïve in the sense that it is both understood and reproduced by non-specialists, by people not associated with or engaged in the practice or community of puppetry. This distinction immediately indicates at least two different perspectives on what a puppet is, which I describe here as 'popular' and 'practitioner', and while my analytical focus is on the former, the analysis also draws out aspects of the latter. To scope this naïve, everyday understanding of the puppet, I focus my analysis on the metaphorical use of puppets in adult fiction and do not seek to forge rigorous links with either disciplinary concepts and concerns or puppetry practices and theories, although disciplinary links do emerge and my initial focus on metaphorical puppets recognises that puppet theatre has been a source of metaphors since ancient times (Jurkowski, 2013). Thus, the focus is on identifying how puppets are represented in fictional metaphors as a route into scoping this quotidian understanding of what a puppet is.

## Chapter summary

Chapter 1 attends to passing metaphorical references to puppets in works of literature that are not nominally about puppets but which use the idea of the puppet to portray aspects of a character or the specific circumstances in which a character finds themselves. The analysis in this chapter is revealing with regard to common representations of different types of puppets and how these make up a fictional puppet appearance and determine the

DOI: 10.4324/9781003214861-3

mobility capacities of the puppetised character. It also gives rise to a specific formulation of a relationship of control between the puppetised character and their puppet master that is focused on the body of the puppetised character. In doing so, it draws out an emphasis on the merging and switching of bodily materiality between wooden puppet and fleshy human to generate graphic and affective impacts on the reader, while simultaneously identifying the eviction of individuality or subjectivity from the puppetised character as a psychological supplement to that corporeal emphasis. The treatment of the border either between puppet and human or between puppet and puppet-master is found to generate four different spatialities, which are distinct yet also overlap in certain ways due to their unique configurations of – for example – bodily versus psychological aspect and materiality versus subjectivity focus.

Chapter 2 progresses to examine narrative metaphorical uses of puppets in works of literature that employ the word 'puppet' in the title but do not feature puppets as material entities in the plot, thereby employing the idea of the puppet as a key narrative device but in metaphorical terms. The analysis in this chapter generates findings that are broadly consistent with those in Chapter 1, for example, in relation to the significance of control and the unsettling of clear borders between puppet and puppet master, but it also diverges in some ways and provides much more detail in others. In terms of divergence, the appearance of specific puppet types does not feature, there is no meaningful engagement with the merging or switching of bodily materials between wooden puppet and fleshy human, and the means of control is predominantly psycho-social rather than bodily. However, in terms of providing more detail, the way in which both the relationship between the puppet and the puppet master is established and the control is exerted are found to be addressed more thoroughly and expressed with greater nuance. Whereas the passing metaphorical uses of puppets emphasised the material interchangeability of puppet and human, the narrative uses of puppets set the relationship of control within a broader social context, facilitating social commentary that is found to link the pathological exercise of control at the level of the individual to the ideological exercise of control at the level of society. This narrative metaphorical use of puppets is also found to engage more rigorously with different ways in which the subjectivity of the puppetised character is evicted, thereby building on the findings of the analysis in Chapter 1, and to reveal various strategies employed by the puppet masters to establish and maintain their control. Three different sets of puppet spatialities – or spatialising processes – are identified in Chapter 2, with one set associated with different strategies of control, another set linked to diverse attitudes adopted by puppet masters to the puppetised characters and the third identified through the consideration of genre influences. Chapter 2, then, reinforces, questions, deepens and diversifies the findings in Chapter 1.

While in this part I engage with geographical literatures with only a light touch, I draw more substantively on the notion of the grotesque, as established in the introduction, due to the affinities between puppets and the grotesque. Puppets are a common theme in the grotesque, but I use this concept as an analytical lens through which to engage with puppets, their bodies and their spaces, which – in turn – generates puppet-informed conceptual contributions to the grotesque. One core feature of the grotesque is the foundational nature of contradictory conceptual categories in the construction of a grotesque form and while such contradictoriness is clearly in evidence in the constructions of puppets explored in Chapters 1 and 2, one of the most telling is the apparent chasm that emerges between these popular cultural constructions of puppets and the understandings of puppets articulated by puppetry practitioners, suggesting that the puppet is not only fundamentally and uniquely grotesque in the forms that it takes but also in the ways in which it is understood, which brings with it implications for what we consider the grotesque to be and how we might work creatively with it.

## Conclusion

The Conclusion to Part 1 summarises and refines the headline outcomes of the first two chapters before elaborating further on the relationship between puppets and the grotesque to consolidate the conceptual contribution of this analysis in developing our understanding of the grotesque. It then specifies the diverse puppet geographies emergent from these analyses and the human–puppet border crossings on which they are predicated and outlines the relevance of these analyses to disciplinary themes and interests, with a focus on the confluence of biopolitics and geopolitics at the level of the individual. Part 1, then, establishes an everyday understanding of what a puppet is as the quotidian baseline for the later analyses, which examine more substantive uses of puppets, for example, as protagonists or actors, and which broaden in focus to other cultural forms, including film and television, music and theatre. At the same time, my disciplinary engagement both builds and deepens as the significance of puppet–human borders and border crossings becomes increasingly clear.

## Reference

Jurkowski, H. (2013) *Aspects of puppet theatre*, 2nd Edition. Palgrave Macmillan: Basingstoke.

# 1 Puppets in passing

In this first substantive chapter, I engage with everyday representations of puppets in popular culture to establish a commonplace understanding of what makes a puppet a puppet. While puppets come in a variety of shapes and sizes, we hold an everyday or naïve understanding of what we mean by the term 'puppet' regardless of the specific form that it may take from one instance to another. Whether we are dealing with a glove or sleeve puppet (operated directly by the hand inside it), a rod puppet (operated using sticks from below), a marionette (operated by strings from above) or any other type of puppet, and whether operated by a single puppeteer or a team of people, there are certain features that we consider to be characteristic of being a puppet. It is these features that constitute the naïve or everyday understanding of what a puppet is, in terms of its appearance, materiality, capacities and relations. Throughout this book, I attend to different ways in which puppets are represented, constructed and used in diverse cultural forms (e.g., fiction, film, music and theatre) and the ways in which these generate their own unique puppet spatialities. Part 1 adopts a principal focus on literary (fictional) representations in works that use puppets metaphorically. Along the way, and further elaborated in the next chapter, I identify a variety of borders – bodily and subjective – emerging from this naïve, everyday understanding of a puppet that contribute to the construction of the puppet in relation to its human counterparts and the generation of their unique spatialities. These various borders and border crossings both provide an immediate link with ideas of the grotesque, as outlined in the Introduction, and will become increasingly focal to the analysis and argument as Parts 2 and 3 unfold.

As a starting point, this chapter explores occasional or passing metaphorical references to puppets in a range of adult fiction, film and television to establish quotidian or everyday assumptions about puppets. While film and television references are included in the analysis, this chapter focuses strongly on passing metaphorical references to puppets in adult fiction, as this is the

DOI: 10.4324/9781003214861-4

dominant source of metaphorical references. None of these works was explicitly about puppets but made metaphorical use of the idea of the puppet to convey either features of the characters involved or the situations in which they found themselves. Approximately 85 passing metaphorical references emerging from 50 works of adult fiction were sourced between 2016 and 2020 for the analysis in this chapter, alongside another 18 from film and television programmes. The full list of these works is presented in Table 1.1 and all works are identified by title irrespective of whether they are literary or filmic in nature. This is done both in the interests of consistency in the treatment of the different types of works and to distinguish the literary works analysed from academic citations. The references to puppets were noted down verbatim either by myself or by family members who have kindly been acting as scouts for me. While this inevitably leads to an emphasis on certain types of cultural output that we gravitate towards, the works listed cover a range of genres (e.g., murder mystery, psychological thrillers, science fiction, action and comedy) and some of the references identified by family members came from works that I would not normally read. In addition, as I am interested in unearthing the core characteristics of a naïve understanding of puppets, this is likely to be reasonably consistent across genres, so a skew towards thrillers and murder mysteries rather than fantasy and romance literature is likely to have minimal impact on the core characteristics identified. This contrasts with the analysis undertaken in the next chapter, which draws upon adult fiction in which the word 'puppet' features in the title, suggesting that the characteristics of everyday understandings of puppets are strong narrative elements in the stories, but in which puppets as material entities do not feature, establishing any puppet references as metaphorical but as a more significant element of the narrative.

The first analysis establishes a baseline understanding of the puppet as crafted in the passing metaphorical use of puppets in adult fiction while Chapter 2 deepens this analysis, exploring in more detail specific aspects of the puppet that emerge from this analysis in the more substantive narrative metaphorical uses of puppets. The Conclusion to Part 1 subsequently reflects on the two discussions to identify commonalities and discrepancies between the different uses of puppets and summarises both the spatialities that emerge from these metaphorical uses of puppets and the various borders – bodily, subjective and more – that contribute to those spatialities. Moreover, the implications of these findings for understandings of the puppet as grotesque and for understandings of the grotesque and especially the body grotesque, are drawn out through consideration of the varied ways in which gaps and borders between different categories (e.g., human/nonhuman subject/object, living/dead, corporeal/psychological) are bridged or overcome in the constructions of puppets encountered.

*Table 1.1* List of works analysed

### Adult Fiction

| | Author and Date | Title | Publisher |
|---|---|---|---|
| 1 | Savage, S. (2010) | The Cry of the Sloth | Orion |
| 2 | Chanter, C. (2015) | The Well | Canongate Books Ltd |
| 3 | Crawford, D. (2011) | Covenant | Simon and Shuster |
| 4 | Rendell, R. (1994) | The Crocodile Bird | Arrow Books |
| 5 | O'Brian, P. (1996) | Master and Commander | Harper Collins |
| 6 | James, P.D. (1988) | Unnatural Causes | Sphere |
| 7 | James, P.D. (2000) | Innocent Blood | Faber and Faber |
| 8 | James, P.D. (2010) | The Black Tower | Faber and Faber |
| 9 | Binchy, M. (1992) | Light a Penny Candle | Arrow Books |
| 10 | Child, L. (1999) | Killing Floor | Bantam Books |
| 11 | Child, L. (2011) | Worth Dying For | Bantam Books |
| 12 | Child, L. (2009) | Nothing to Lose | Bantam Books |
| 13 | Child, L. (2006) | One Shot | Bantam Books |
| 14 | Child, L. (2010) | 61 Hours | Bantam Books |
| 15 | Child, L. (2018) | Past Tense | Bantam Books |
| 16 | Follett, K. (2015) | Edge of Eternity | Pan Books |
| 17 | Patterson, J. and Fox, K. (2016) | Private Sydney | Arrow Books |
| 18 | Booth, S. (2017) | Secrets of Death | Sphere |
| 19 | Harris, R. (2016) | Dictator | Arrow |
| 20 | Blackwood, G. (2016) | Tom Clancy's Duty and Honour | Penguin |
| 21 | Swift, G. (2011) | Wish You Were Here | Picador |
| 22 | Rhys, R. (2017) | Dangerous Crossing | Doubleday |
| 23 | Cornwell, P. (2017) | Chaos | Harper Collins |
| 24 | Bear, G. (1987) | The Forge of God | Arrow Books |
| 25 | Reichs, K. (2004) | Bare Bones | Arrow Books |
| 26 | Süskind, P. (1986) | Perfume | Penguin |
| 27 | Stallwood, V. (1994) | Oxford Exit | Headline |
| 28 | O'Brien, E. (2015) | The Little Red Chairs | Faber and Faber |
| 29 | Tchaikovsky, A. (2016) | Children of Time | Pan Books |
| 30 | George, E. (1992) | For the Sake of Elena | Bantam Books |
| 31 | Christie, A. (1989) | The Underdog | Fontana |
| 32 | Holt, V. (1978) | The Pride of the Peacock | Fontana |
| 33 | Smith, W. (2014) | The Seventh Scroll | Pan Books |
| 34 | Smith, W. (2013) | Monsoon | Pan Books |
| 35 | Smith, W. (2017) | War Cry | Harper Collins |
| 36 | Smith, W. (2017) | The Tiger's Prey | Harper Collins |
| 37 | Francis, D. (1992) | Comeback | Pan Books |
| 38 | Weaver, T. (2013) | Never Coming Back | Penguin |
| 39 | Patterson, J. and Paetro, M. (2017) | 16th Seduction | Arrow Books |
| 40 | Hawkins, P. (2016) | The Girl on the Train | Black Swan |
| 41 | Sheldon, S. (1991) | The Doomsday Conspiracy | Harper Collins |
| 42 | Clarke, A.C. (1970) | The Deep Range | Pan Books |
| 43 | O'Brian, P. (2007) | The Mauritius Command | Harper Perennial |

*(Continued)*

*Table 1.1* (Continued)

**Adult Fiction**

| | Author and Date | Title | Publisher |
|---|---|---|---|
| 44 | Abbott, M. (2014) | The Fever | Picador |
| 45 | Patterson, J. and Karp, M. (2018) | NYPD Red 5 | Arrow Books |
| 46 | Smith, T.R. (2011) | Agent 6 | Shuster and Shuster |
| 47 | Brown, D. (2002) | Deception Point | Corgi Books |
| 48 | Crichton, M. (1994) | Disclosure | Arrow Books |
| 49 | Wiggs, S. (2017) | Map of the Heart | Harper Collins |
| 50 | Bagley, D. (1984) | Night of Error | Brockhurst Publications Ltd |

**Film**

| | Title | Director | Year |
|---|---|---|---|
| 1 | The Husband She Met Online | C. Crawford | 2013 |
| 2 | Buried Secrets | M. Mitchell | 2015 |
| 3 | Murder at 1600 | D.H. Little | 1997 |
| 4 | Hot Fuzz | E. Wright | 2007 |
| 5 | Shark Tale | R. Letterman, V. Jenson, B. Bergeron | 2004 |
| 6 | Sully | C. Eastwood | 2016 |
| 7 | Dr Dolittle | R. Fleischer | 1967 |
| 8 | Mrs Doubtfire | C. Columbus | 1993 |
| 9 | Mindhunters | R. Harlin | 2004 |

**Television**

| | Series | Episode | Date/Year |
|---|---|---|---|
| 1 | Star Trek: The Next Generation | Shades of Gray | 1989 |
| 2 | Star Trek: The Next Generation | Measure of a Man | 1989 |
| 3 | Diagnosis Murder | A Candidate for Murder | 1996 |
| 4 | Diagnosis Murder | Physician, Heal Thyself | 1997 |
| 5 | Diagnosis Murder | A Resting Place | 2000 |
| 6 | The Jeremy Vine Show | – | 05/09/19 |
| 7 | Strictly Come Dancing | – | 13/10/18 |
| 8 | Strictly Come Dancing | – | 20/10/18 |
| 9 | Funniest Ever TV Cock-Ups | – | 09/06/19 |

## A naïve understanding of puppets

In the 68 works analysed that make passing metaphorical use of puppets, the term puppet is employed in literature to refer to a general type of manipulable object, whereas in film and television passing metaphorical reference is made to puppets both in this generic sense and in relation to specific puppet characters. This is most commonly Pinocchio, as in the portrayal of the android Data as a futuristic Pinocchio – in wanting to become a real person

– in *Star Trek: The Next Generation*. Marionettes (string puppets) are by far the most common form of puppet employed, and key features of puppets that attract comment include their size, their faces, their movement (or lack thereof) and the idea of being controlled. Each of these is addressed in turn, under headings of visuality, mobility and control, before the spatialities of these metaphorical uses of puppets are identified and the role of borders in the constitution of these puppet spatialities is unearthed.

### *Visuality*

Visual features of puppets that emerge from these works revolve around the size and facial features of the puppet, which lead to the dominance of a specific type of puppet in visual representations. Addressing size first, puppets are commonly assumed to be small, which is achieved either through the use of language to convey a perception of somebody being reduced in size or through contextualisation: 'dwarfed as puppets' (*Unnatural Causes*, p101). Only two works reviewed talk about puppets being large and one of these could just as easily be interpreted as presenting puppets as small through comparison with the scale of the human. Swift (2011) writes of a character 'moving around like a big, jerky puppet' (*Wish You Were Here*, p198), which – rather than suggesting that puppets are large – could be seen to suggest that puppets are inherently smaller than humans if a human moving in a jerky fashion is an oversized puppet. Puppets, then, are predominantly diminutive in stature, capable of being packed up in a box (*The Crocodile Bird*) or tucked into a pocket (*16th Seduction*), but this emphasis on the small is notable in contrast to the diverse sizes of puppets that are commonly seen in puppetry practice, with recent examples including the life-size horses in theatrical productions of *War Horse* and the larger-than-life 'giant' puppets that periodically grace the UK's streets, from the 37-ft-high *Man Engine* in Cornwall (Anon 2018) to the 33-ft-tall *Storm* in Glasgow (Anon 2020). Despite this apparent gulf between naïve understandings of puppets in popular fiction and the realities of contemporary puppet practice, even these metaphorical literary puppets might find themselves caught in a web of extensive relations of exploitation (*The Doomsday Conspiracy*), thereby introducing a scaling effect and a distributed network of control, extending the characteristic puppetness beyond the bodily form of the puppetised character itself. However, and conversely, this distributed network of puppetisation further reinforces the diminutive nature of these metaphorical puppets as they are just one element within the broader network, rendering them as both small in stature and dwarfed by context.

Descriptions of a puppet's appearance focus primarily on its face and it is in relation to faciality that ventriloquists' dummies rather than marionettes attract the greatest attention. Puppets are portrayed as being highly unnatural in appearance, with exaggerated lines and grotesque features, including a foreshortened face, a short bony nose and a shapeless void of a mouth (*The Black Tower*). One character is described as having 'deep lines scored down each side of her mouth' (*Dangerous Crossing*, p207), while another had a tight chin, as if it had been 'pulled taught' by wires (*The Fever*, p148), both of which go on to draw a direct comparison with ventriloquists' dummies. In both cases, the reference to a ventriloquist's dummy evokes stronger visual and affective associations than the description of the character's actual facial features, which is neatly encapsulated in one character's consideration of another's appearance: 'like the ventriloquist's dummy that used to perform at the mall, Deenie thought, then felt bad about it' (*The Fever*). Whereas the tightness of the chin was a straightforward description, its association with a ventriloquist's dummy introduced a normative dimension and an affective response as resembling a ventriloquist's dummy is to be avoided. This is the first and very clear instantiation of the grotesque within these constructions of puppets, not only through the embodied combination of the fleshiness of the human face with the facial characteristics of a ventriloquist's dummy, but also in the affective juxtaposition of playfulness and disgust evidenced in the sudden shift to remorse on the part of Deenie despite reminiscing on a playful encounter with a puppet at the mall in recognition that the association that has just been drawn is inappropriate. These two key aspects of the grotesque – the combination of contradictory categories and the mix of playful and disgust or horror (Kayser 1963; Harpham 1976, 2006; Connelly 2012; Pilný 2016) – are clearly in evidence here, and, while not universal, the strong association of facial appearance with a ventriloquist's dummy highlights the assumption of a very stylised appreciation of puppets among the anticipated readership. Moreover, the combination of a ventriloquist dummy's head on the body of a marionette constitutes its own body grotesque, not through the merging of human and puppet forms but through the merging of two distinct puppet types.

## *Mobility*

Along with the face, other body parts that receive attention are arms and legs, but this is almost exclusively in the context of movement, and in relation to which the metaphorical attention turns to marionettes. The most common term for this movement is jerking (*The Pride of the Peacock*; *Oxford Exit*; *Wish You Were Here*; *The Girl on the Train*), although others include flapping, flailing, twitching, kicking and dancing comically. Such descriptors evoke erraticism, a lack of control and extremism as people are jerked and tossed around against their will, controlled by strings in another grotesque combination of the comic and the horrifying (Kayser 1963; Harpham 1976, 2006). The material contrast between string and flesh in the portrayal of humans being controlled as if they were marionettes provides another combination of contradictory categories (this time human and non-human materiality) and further reinforces the assumed diminutive scale of the puppet, but materiality also emerges in another sense in the context of mobility by associating the puppet's woodenness with stiffness or awkwardness. Characters either 'tottered away as if on wooden legs' (*Perfume*, p26) or bobbed 'like a puppet on a stick' (*Deception Point*, p540). This emphasis on wood, too, is at odds with contemporary puppet making and practice, which employs diverse materials as well as forms and scales (Tillis 1996; Francis 2012), from papier mâché and fabric to metal, foam and clay, and which is valued for manipulative prowess and gracefulness of movement (Leach 1983; Francis 2012). Similar emphasis is found in critical reviews of puppet performances, with *The First Hippo on the Moon* among others drawing criticism for sloppy execution or rudimentary puppetry or expressiveness (Cavendish 2015; Taylor 2015; Cavendish 2016a; Cavendish 2016b; Lee 2017; Mountford 2019), and the puppet productions securing the most favourable reviews being those deemed to deliver the highest quality puppetry skill. *The Lorax* was praised for the artful manipulation and voicing of the puppets, which was judged to have been done with dignity and dexterity (Billington 2015; Taylor 2015), while *War Horse* is now firmly established as an iconic cultural work and exceptional puppetry practice, consistently described as awe-inspiring and incredibly realistic (Herring 2019), touchingly expressive and remarkably articulate (Hitchings 2018) and even as 'the most plausible and expressive quadrupeds ever to have graced the London stage' (Billington 2009, n.p.). There is clear contrast, then, between the movement capacities of metaphorical puppets and the movement capacities of puppet performance that are appreciated by practitioners and critics alike. What is unclear here, though, is whether the quality of the character's mobility is suggestive of woodenness or whether the assumed woodenness of the puppet bodily form gives rise to the quality of movement. Either way, the puppetised character is controlled from without in a combination of metaphorically wooden form and stringy manipulation that merges human fleshiness and objectual materiality and constructs the puppet as a specific hybrid form (a ventriloquist's dummy's head on a marionette's body) and a specific materiality (wooden), giving rise to a characteristic mobility.

Contrasting with this jerky erraticism is a common depiction of a lack or sudden loss of movement, introducing a binary to the mobility of the puppet. Collapsing (*The Forge of God*; *Children of Time*), slumping (*The Black Tower*) and crumpling (*Killing Floor*) all denote this sudden change in mobility, which is metaphorically described in puppet terms as being unstrung, released from strings – or most commonly – as if the strings had been cut (*Killing Floor*, *One Shot*; *Nothing to Lose*; *Agent 6*), neatly and simultaneously encapsulating the suddenness and surprise associated with the grotesque (Kayser 1963) and the centrality of the strings in facilitating the mobility of the puppetised human character. Notably, this sudden cutting or removal of the strings by which the puppetised character gained its erratic movement does not return the puppetised character to its former human status but simply renders it inert, an object. Seemingly, then, the puppet is characterised both by a specific – jerky, twitchy – style of movement and a total and sudden loss of movement, while the character's agency is permanently removed once they have been puppetised, even if the means of control – the strings – are severed. This grotesque merging of human and puppet is seemingly one-way: the human becomes puppetised but there is no puppet being humanised, which would or might gain its own motoric agency once the strings are cut.

So, although my analytical emphasis is on the puppets, the employment of the grotesque in these works of fiction is concerned – in typical grotesque fashion – with questions of what it is to be human. It is the human that is changed, and it is therefore what it is to be human that is questioned.

## Control

The strings, then, are the metaphorical means by which the jerky movement of the puppetised character is controlled as they are forced to do things at the will of another, like a plaything, which introduces another binary between playfulness and brutality, and one that lies at the heart of the grotesque as tragi-comic (Kayser 1963; Harpham 1976, 2006; Connelly 2012). The notion of play itself is significant here, in two senses of the word, as in comments about characters 'being played' like puppets (*Diagnosis Murder – Physician Heal Thyself; Buried Secrets; Murder at 1600*) the meaning of play conveyed for the character being controlled is not about personal enjoyment but about their individual triviality and lack of agency, while for the character acting as a puppet master, it can be very much about the pleasure derived from cruelty. Literature on the grotesque acknowledges the combination of the playful and the horrific in such aesthetic forms and some authors emphasise that play is the grotesque's most pervasive characteristic (Connelly 2012), but these metaphorical constructions of puppets make clear that what constitutes playfulness for the puppet master inflicts horrific cruelty on the puppetised character: The two are not simply experienced side by side but coincide in the same event. Moreover, they remind us that playfulness itself carries more than one meaning, with one of those meanings evoking demeaning trivialisation rather than light-hearted entertainment, which itself emphasises the contradiction between the powerfulness of the puppet master and the powerlessness of the puppetised character. While many of the metaphorical references to puppets portray characteristics of the puppetised character, when a puppet master is brought into the scene, they are commonly described as ruthless (*The Doomsday Conspiracy*), diabolical (*Chaos*) or mad (*War Cry*), either tormenting one puppetised character or setting one against another to further their own ends (*Tiger's Prey*) in a display of the psychological excess of the grotesque (Duggan 2016). The contrast emerging between the enjoyment of the puppeteer and the distress of the puppetised character serves to enhance the affective power of the predicament in which the puppet finds itself. With this binary, the sense of playfulness contrasts with the presumed nastiness of the puppeteer in classic grotesquery, while also emphasising that nastiness by rooting that playful pleasure in the purposeful brutalisation of the puppet.

The strings of a marionette are key to the relationality depicted in these works, and this relation is condensed into one of control. Significantly, this control is enacted very much at a bodily level, as evidenced above in the twitching and flailing of the puppetised bodies and the sudden absence of mobility when the puppet bodies slump. While brutality inflicted on the body also shows through in a more immediate sense through references to bloodstains, twisted necks and disconnected joints (*The Black Tower; The Well; War Cry*), again combining the fleshy vulnerability of the human with the material objectification of the puppet, control of the puppet's body and actions is commonly generalised to the jerkiness of movement and the sudden loss of that movement. Much less in evidence is any psychological or emotional form of control explicitly related to the notion of the puppet, further reinforcing the objectification of the human. There is some evidence of this, in the form of fear as to what the puppet master will do, but this is itself grounded in the threat of physical bodily harm or destruction, as when one puppetised character is faced with trading his own life for that of his sister (*One Shot*). Only one of these works explicitly addresses the effectiveness of psychological control compared with physical threat, proposing that it is more effective to turn a child's mind against its parents than to kidnap and threaten the child (*Tom Clancy's Duty and Honour*). In such psychologised narratives, the web of exploitation is widened, as in the example of the child, both the child and its parents are puppetised, with the child being controlled in order to control the actions of the parents. While these occasional instances of psychological control provide for a more distributed network of control than direct threats by stringing together webs of familial and social contacts that can be manipulated, it is the body of the puppet – and harm (or damage) inflicted thereupon – that is the dominant focus for the control of the puppetised character in these works.

The relation between a puppet and a puppet master is not just one of brutal control, however, as the puppet is not only dehumanised and objectified, but is also rendered simple and unintelligent compared with the puppet master who makes plans (*One Shot*) and masterminds coups (*The Doomsday Conspiracy*). In another parallel with the grotesque whereby modification of bodily form is associated with intellectual deviancy (Harpham 1976), the puppet master is deemed to possess greater intelligence (*Worth Dying For*), with those that they control being intellectually deficient: 'there seemed to be a minimum of brain-power around – these men were mostly obedient puppets' (*Night of Error*). The idea of being 'played with' features again here, as the character being controlled is infantilised as a toy for the amusement of the character who directs the 'game'. Further, the inability of the puppetised character to direct their own actions is described in terms of a loss of self, identity or subjectivity: They are not themselves anymore, not in conscious control of their actions (*Wish You Were Here*). In such uses of the puppet, the individuality of the character is evicted – disembodied – by the intellectually superior puppet master at the very time that the body of the puppetised entity takes centre stage, in a hollowing out of the metaphysical persona from the physical confines of its body. Again, there is a contrast here between the lack of agency attributed to the fictional puppetised character and the sense among puppetry professionals that puppets do in fact do their own thing and cannot be fully controlled (despite lacking any persona or psychological self) and the need for puppeteers to learn to adapt their behaviours to suit the demands of the puppet (Bicât 2007; Zamir 2010; Gross 2011).

Seemingly, then, the reader is asked to empathise with the puppetised character even though their character has been evicted, while the brutality inflicted on the body of the puppetised character is undermined by the material associations of the puppet with wood even as it is used to stimulate the required empathy. Ultimately, the affective power of these metaphorical uses of puppets lies in the blurring and crossing of borders and distinctions between puppet and human, in relation to both body and subjectivity, leading to a contradictory situation in which the reader is enticed to empathise with an objectified character as the very objectification of the character simultaneously allows for enhanced brutality and dehumanisation and erases the character with whom we are encouraged to empathise. We empathise with the eradicated humanity of the puppetised character that has been stripped of its mobility, agency and subjectivity, despite there being no humanity left with which to empathise, and we empathise with the brutality inflicted on the puppetised character even though their puppetisation evokes woodenness and thereby nullifies the risk of pain or death. This is a profoundly grotesque contradiction, grounded in the blurring of human/puppet distinctions, from which the reader is challenged to forge meaning and affective identification with the puppetised character by walking a fine line between progressive objectification and sustaining subjectification. The objectification is necessary to enhance the sense of brutality and powerlessness on the part of the puppet but sufficient subjectification must be maintained in order for the reader to empathise with the puppetised character's plight. Even in this most everyday understanding of a puppet constructed through passing metaphorical references, a contrariness of puppets is consistent with the contradictoriness of the grotesque and is used for powerful affect – by juxtaposing jerky movement and loss of mobility, the physical forms of a ventriloquist's dummy and a marionette, playfulness and brutality, subject and object – and these juxtapositions generate notable puppet spatialities through the creative muddling of puppet–human borders.

## Puppet spatialities

The contrariness of puppets teased out through the preceding discussion was established through the combination of contradictory categories and the creative muddling of several borders. Distinct from the notion of the grotesque either lying in the gaps between these contradictory categories or incorporating multiple categories (Harpham 2006), these grotesque puppet forms are generated through forging bridges across those gaps and between those categories: Turning the gaps into generative borders that are reshaped through varied and ambiguous interpenetration of the multiple categories involved. While still amenable to consideration as a construction of the between or the liminal (Kayser 1963; Pilný 2016), being neither one thing nor the other yet both at the same time (van Gennep 2010), the significance of thinking in terms of connections rather than gaps is that it draws attention to how those connections are forged to enable the reader to walk that fine line between objectification and subjectification, and to enable them to empathise with a character that has been evicted from the bodily form that is the object of extreme brutality. Rather than being a concept without a form (Harpham 2006), this is more a case of a form that cannot easily be conceptualised. The general form of the metaphorical puppet was found to be a composite of the face of a ventriloquist's dummy and the body and mobility of a marionette, merging two puppet forms that would normally be discrete and generating both a comingled metaphorical space of the puppet body and a distinctive visuality in terms of both appearance and stylised movement, which is associated with a specific materiality (woodenness). This establishes a distinct configuration of the naïve puppet's body grotesque, not solely between human and object but initially between different puppet forms, which are then merged with a human character.

The shift in analytical focus from appearance to movement drew attention to the materialities at play in these metaphorical puppet constructions, whereby the presupposed woodenness of the puppet and the fleshiness of the human are interwoven and transposed for graphic and affective purposes. The brutality that can be inflicted on the puppetised character is magnified by the puppetisation that grants them wooden rather than fleshy form but that puppetisation – in rendering the character wooden rather than fleshy – simultaneously undermines the affective power of the brutality that is inflicted. The focus on control through bodily harm, then, is counteracted by the shift in bodily materiality from flesh to wood through the metaphorical establishment of the character as a puppet, yet the reader empathises with the puppetised character despite this confused materiality, generating a counterintuitive affective space of material transposition.

At the same time, although the use of psychological control is far less evident in these works than physical force, there is a certain psychologised aspect to these metaphorical uses of puppets in the diminishment of intellect on the part of the puppetised character compared to the intellectually superior puppet master. This reinforces the lack of agency on the part of the puppetised character arising from the physical control through metaphorical strings and from the lack of agency even after those metaphorical strings have been cut. The individuality, intellectual capacity and agency of the puppetised character are evicted from the bodily form that is controlled and brutalised by the puppet master, even while that eviction process is used to stimulate empathy and sympathy on the part of the reader for that character that has been evicted. This psychological aspect, then, generates a vacant space of subjectivity that – nonetheless – evokes empathy and sympathy on the part of the reader.

Finally, the presence – albeit limited – of psychological means of control in these works allows for puppetised characters to be subjected to control and exploitation through extended networks of contacts, which in turn stretches and multiplies the metaphorical strings through which that control is exerted and reinserts the subjectivity of the character through their susceptibility to manipulation through threats to those about whom they care. In contrast to the tightly body-bound comingling of puppet types in one physical form, and reflecting the unfinished, social nature of the grotesque body (Bakhtin 1984; Russo 1994; Li 2009), this is an expansive distributed space of subjective non-agency that – where present – supplements the affective power generated by creative comingling of categories across the human–puppet bodily border.

Overall, this analysis has identified four spatialities of metaphorical puppet that differ in terms of aspect (bodily versus psychological), focus (form, materiality or subjectivity) and process (merging, transposing, evicting, extending), giving rise to the metaphorical puppet being characterised on a continuum from object to human. This continuum is shown diagrammatically in Figure 1.1.

In spatialities 1 and 2, the bodily aspect of the puppetised character is primary and in general terms, the character is closer to object than human, but spatiality 1 is generated by merging different puppet forms (marionette and ventriloquist's dummy) while spatiality 2 is created through the transposition of puppet and human materialities (flesh/wood). In spatialities 3 and 4, the psychological aspect is primary and in general terms, the character is closer to human than object, but spatiality 3 arises through the eviction of the puppetised character's subjectivity (hollowing out) whereas spatiality 4 reflects the extension of the puppet-form through the networked space of control (distanciation). Reinserting the puppetised character's subjectivity in order to manipulate them via their familial and social relations means that its status in spatiality 4 is closest to human, although the extensive network of control precludes any agency.

The puppetised character at the heart of spatiality 2 is classified as a humanised object as the materiality of the puppet is paramount to both the brutality inflicted and the challenge of empathising with the character, while that at the heart of spatiality 3 is classified as an objectified human as it is the absence of subjectivity that distinguishes the puppetised character in spatiality 3 from spatiality 4. Importantly, the affective power of these

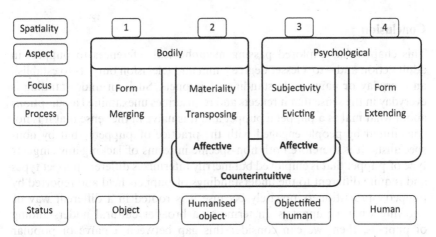

*Figure 1.1* Metaphorical puppet spatialities.

Source: Author.

metaphorical puppets is greatest in spatialities 2 and 3, whereby the border between the object (puppet) and the subject (human) is most muddied, and is greater still if spatialities 2 and 3 coalesce such that both the materiality and subjectivity of the puppetised character are confused or contradictory. Notably, and as outlined earlier, this affective power seems counterintuitive precisely because of the muddling of materiality and/or subjectivity, as readers are induced to empathise with a character who has been dehumanised in the very process of puppetisation that generates its unique spatiality and its affective power. This most extreme version of the grotesque puppet body is not simply binary but is contradictory along potentially multiple axes as connections and interpenetrations between incompatible categories of existence are brought into play to walk a fine line between the objectification and subjectification of the puppetised character, which – if successful – perversely induces empathy for a character that has been reformulated as a marionette-dummy hybrid, rematerialised and differentially mobilised as wooden, and has had their subjectivity simultaneously hollowed out and distanciated.

While the conceptual categories that constitute the contradictions from which these puppets are constructed remain isolated from each other, the form of the metaphorical puppet integrates those categories in diverse ways, which both creates varied spatialities of the puppet and prompts a rethinking of the grotesque. Here, the grotesque is not considered as a concept without form (Harpham 2006) but as a form that challenges conceptualisation, not solely as a space of social or political potential (Bakhtin 1984; Li 2009; Connelly 2012; Duggan 2016; Pilný 2016) but as variously spatialised, and not as lying either in multiple categories or in the gaps between categories (Kayser 1963; Harpham 2006; Pilný 2016) but in the reshaping of the borders between categories and the forging of creative connections across those categories, bridging the gaps in novel ways and creating unique puppet formulations and spatialities.

## Conclusion

This chapter has explored passing metaphorical references to puppets in adult fiction and – to a lesser degree – film and television outputs to establish an everyday or naïve understanding of puppets. Such an understanding is everyday in the sense that it reflects and re-inscribes unexamined assumptions about what makes a puppet a puppet and it is naïve in the sense that it is not constituted by people engaged with the practice of puppetry but by non-specialists. It is therefore both non-specific in terms of lacking any singular type of puppet that is employed but merrily intermixes different puppet types and is quite different to the understandings of puppets held and reported by puppetry practitioners precisely because it is rooted in a different way of knowing what a puppet is. In terms of a broader cultural understanding of puppets, then, we can consider this gap between a naïve or popular understanding and a practitioner understanding of puppets to be just as constitutive as the conceptual gaps identified above are constitutive of the naïve constructions of puppets. Any reader of one of these fictional metaphorical engagements with puppets who has been exposed to contemporary puppet practice must find a way to accommodate both these understandings – popular and practitioner – in their reception of and response to the fiction, recrafting the borders between the two understandings.

In different ways and to different degrees, the transposability of wooden materiality and fleshy corporeality is used creatively for affective purposes, while the connectivity of the strings establishes the mechanism of control. Consequently, we can discern four distinct puppet spatialities that emerge from this analysis, which were conceptualised as a continuum of overlapping aspects and processes. The first is a comingled or merged space of two puppet forms (ventriloquist's dummy and marionette). The second is a bodily space of transposed materialities between puppet and human, using the materiality of the puppet to highlight the objectification of the human character and the fleshiness of the human to evoke empathy through the brutality inflicted on the puppet. The third is a vacant space of subjectivity, generated by hollowing out the body of any individuality or evicting the puppetised character's subjectivity. The fourth is a networked space of relations, which extends the reach of the controlling strings and reinserts the puppetised character's subjectivity while still precluding its agency.

Throughout the discussion, the relevance of the grotesque to constructions and understandings of puppets has been kept in mind, and both consistencies and discrepancies between the grotesque and puppets have emerged. For example, the combination of contradictory categories and the juxtaposition of the playful and the horrific that are so characteristic of the grotesque were clearly in evidence in these fictional constructions of puppets, but were extended to the consideration of the construction of a puppet's body grotesque through combining two different types of puppet before merging the composite puppet with the human being puppetised and to a nuanced consideration of different ways in which the playful and the horrific can be not only juxtaposed but coincidental. Moreover, this exploration of passing metaphorical references to puppets has prompted a rethinking of certain aspects of the grotesque, specifically a reversal in its relationship between form and concept (from a concept without form to a form that challenges conceptualisation), a reformulation of the nature and significance of the gaps between contradictory categories (from lying in the gaps between categories to forging bridges across those gaps by recrafting the borders between them) and a shift in spatial emphasis (from the grotesque as a space to the varied spatialities of diverse grotesques). Seemingly, then, the grotesque can certainly be fertile conceptual terrain for geographical engagement with puppets and the body grotesque provides a specific lens through which to explore the treatment of borders between puppet and human bodies beyond that of the puppet and puppeteer, but the puppet can also be fertile analytical territory for a reconceptualisation of the grotesque and geography's sensitivity to the spatial can provide an equally productive analytical lens for both puppets and the grotesque.

Overall, the puppet as constructed in these works is something of a mish-mash, but with strong visualities and mobilities, affective power and scaling effects, lending the puppet to further geographical analysis. Firstly, this fictional puppet is a combination of facial features from a ventriloquist's dummy and the body and mobility of a marionette, establishing this metaphorical puppet as a composite of two material forms of puppet and forging links to two core concerns for contemporary geography (visuality and mobility) as well as current interest in composite entities and concepts (e.g., hybridity). Second, the woodenness of its bodily capacities is emphasised in terms of its movement potential but disavowed in the central focus on the body as the locus of control and brutality which relies on the fleshy ability to feel pain for its graphic effect, forging further links with disciplinary interests in the body, affect and more-than-human or socio-material assemblages. Finally, the individuality of the puppetised character is simultaneously evicted by the agenda-driven control of the intellectually superior puppet master and foregrounded by the demand placed on the reader to empathise with the character and sympathise with their predicament, establishing yet further relevance to geographical themes of subjectivity and (post)humanism. The Conclusion to Part 1 revisits these disciplinary threads to develop these potential contributions further, focusing on the eviction of subjectivity and scaling effects due to their continued significance in Chapter 2, and by integrating these disciplinary threads with those that emerge from the next chapter, discusses them in the context of geography's interest in social commentary and critique.

Evidently, the blurring, crossing and reconfiguring of borders between human and puppet bodies – even in only passing metaphorical references to puppets – allow for the creation of powerful affects and specific puppet spatialities, suggesting a cultural form and practice ripe for geographical and broader social scientific engagement, and for further conceptual, critical and creative engagements with the grotesque in its many and varied forms. In the next chapter, I explore what further insights might be gleaned from literary works that engage more fully and more directly with the idea of the puppet in a deeper interrogation of the narrative use of puppets as a titular focus of a novel.

## References

Anon. (2018) Master puppet (photograph label headline). *Daily Telegraph*, 31 March 2018, p. 14.

Anon. (2020) Learning the ropes (photograph label headline). *Sunday Telegraph*, 19 January 2020, n.p.

Bakhtin, M. (1984) *Rabelais and his world*. Indiana University Press: Bloomington, IN.

Bicât, T. (2007) *Puppets and performing objects: A practical guide*. The Crowood Press: Marlborough.

Billington, M. (2009) War Horse. *The Guardian*, 06 April 2009. https://www.theguardian.com/stage/2009/apr/05/theatre-review-war-horse. Accessed 18 October 2020.

Billington, M. (2015) Dr Seuss's The Lorax review - the best family show since Matilda. *The Guardian*, 17 December 2015. https://www.theguardian.com/stage/2015/dec/17/dr-seuss-the-lorax-old-vic-london-family-show-david-greig. Accessed 18 October 2020.

Cavendish, D. (2015) Dr Seuss hits the stage in a lightweight caper. *The Daily Telegraph*, 17 December 2015, n.p.

Cavendish, D. (2016a) Walliams's hippo shoots for the moon. *The Daily Telegraph*, 20 December 2016, p. 24.

Cavendish, D. (2016b) Potter's rebellious Peter Rabbit dares to seize the day. *The Daily Telegraph*, 29 June 2016, n.p.

Connelly, F.S. (2012) *The grotesque in western art and culture: The image at play*. Cambridge University Press: New York

Duggan, R. (2016) *The grotesque in contemporary British fiction*. Manchester University Press: Manchester.

Francis, P. (2012) *Puppetry: A reader in theatre practice*. Palgrave Macmillan: Basingstoke.

Van Gennep (2010) *The rites of passage*. Routledge: Abingdon.

Gross, K. (2011) *Puppet: An essay on uncanny life*. University of Chicago Press: Chicago, IL.

Harpham, G. (1976) The grotesque: First principles. *The Journal of Aesthetics and Art Criticism*, 34 (4): 461–468.

Harpham, G. (2006) *On the grotesque: Strategies of contradiction in art and literature*. The Davies Group Publishers: Aurora, CO.

Herring, N. (2019) Getting in the saddle for an emotional gallop. *Oxford Times*, 29 August 2019, p. 34.

Hitchings, H. (2018) War Horse Review: Heart-tugging, visually inventive spectacle is back in the saddle. *Evening standard*, 09 November 2018. https://www.standard.co.uk/go/london/theatre/war-horse-national-theatre-review-a3985721.html. Accessed 18 October 2020.

Kayser, W. (1963) *The grotesque in art and literature*. Indiana University Press: Bloomington, IN.

Leach, R. (1983) Punch and Judy and the oral tradition. *Folklore*, 94(1): 75–85.

Lee, C. (2017) Theatre Review: The First Hippo on the Moon. *medium.com/Artmag*, 06 March 2017. https://medium.com/artmagazine/theatre-review-the-first-hippo-on-the-moon-les-petits-f032e944ecdd. Accessed: 20 October 2020.

Li, M.O. (2009) *Ambiguous bodies: Reading the grotesque in Japanese Setsuwa Tales*. Stanford University Press: Stanford, CA.

Mountford, F. (2019) Where is Peter Rabbit? Review: How does a play based on Beatrix Potter's books go so wrong? *Evening Standard*, 11 April 2019. https://www.standard.co.uk/go/london/theatre/where-is-peter-rabbit-review-theatre-royal-haymarket-a4115006.html. Accessed 20 October 2020.

Pilný, O. (2016) *The grotesque in contemporary Anglophone drama*. Palgrave Macmillan: London.

Russo, M. (1994) *The female grotesque: Risk, excess and modernity*. Routledge: New York.

Taylor, P. (2015) The Lorax, Old Vic, theatre review: This is terrifically inventive. *The Independent*, 18 December 2015. https://www.independent.co.uk/arts-entertainment/theatre-dance/reviews/lorax-old-vic-theatre-review-terrifically-inventive-a6778971.html. Accessed 20 October 2020.

Tillis, S. (1996) The actor occluded: Puppet theatre and acting theory. *Theatre Topics*, 6(2): 109–119.

Zamir, T. (2010) Puppets. *Critical Inquiry*, 36(3): 386–409.

# 2 Narrative puppets

This second stage of analysis attends to more substantive uses of puppets for narrative purposes, but still in a metaphorical capacity. In the texts considered here, puppets and/or their associated puppet masters are employed within the title of the work, investing the whole narrative with a thematic emphasis on the nature of a puppet. Given the similarities evident in the titles of these works, they are identified by reference to their author's surname/s. The initial thematic analysis of the seven works considered in this section revealed a lot of consistency but also a degree of variation between them, as outlined in Table 2.1. The analysis in this chapter focuses on the first six of these works as the seventh, although displaying similar themes to the others, is sufficiently different in how these themes are employed to be used as a counterexample to test the robustness of the core analyses, which I explore towards the end of this chapter.

Identified in each of the works surveyed were themes of control, justice/revenge and the relation between a puppet and its operator (master). While all the works emphasised situations of control, the means by which this control was established, as well as its purpose and directness, varied between the works. Relatedly, justice and revenge are featured in all the works, often as the reason for the orchestration of control, but the relation between revenge and justice – whether the same thing or distinct – was constructed in different ways. The relation between a puppet and its operator was similarly constructed in different ways, in some cases evoking similarity, complementarity or personality-based connections between the two entities and varying in terms of whether this relation was framed as, for example, creation, exploitation or domination. Identified in nearly every work were characteristics associated with horror, including darkness of place or human spirit, monstrous elements and the incorporation of spirit possession. Slightly less frequent but still occurring in more than half the novels reviewed are themes of secrecy/deception (e.g., manipulation of knowledge) and entertainment (e.g., control being deemed fun for the puppet master). Questions surrounding a puppet's vitality/objectivity were not always directly addressed, but in many ways, these are implicit in the relation between the puppet and its master, as is the establishment of opposites in the construction of this relationship. This leaves progressions – for example, in terms of severity of act or force of

DOI: 10.4324/9781003214861-5

*Table 2.1* Main themes identified in narrative use of puppets

| Work<br><br>Theme | Craven 2019 | Lewis-Thompson 2019 | Brown and DeFelice 2016 | Heinlein 1951/2010 | Osborne 2017 | Redmond 2000 | Ramsay 2016 |
|---|---|---|---|---|---|---|---|
| Control | X | X | X | X | X | X | X |
| Justice/revenge | X | X | X | X | X | X | X |
| Horror | X | X | | X | X | X | X |
| Knowledge/ secrets | X | | | X | X | X | |
| Puppet/master relations | X | X | X | X | X | X | X |
| Progressions | | X | | | | X | X |
| Entertainment | | X | | X | X | X | |
| Opposites | X | | | | X | X | X |
| Vitality/ objectivity | | X | X | X | | | X |

instruction – as the least frequently occurring theme identified. However, as these are consistent with the findings of the analysis in Chapter 1 and they establish a broader social and explicitly scalar perspective on the matter of control, they are retained as focal points for discussion. Consequently, this discussion is oriented around two overarching themes that emerged from the previous chapter, but which are developed much more fully in the six works considered here: (1) Control, which sweeps up issues of revenge/justice and leads to a consideration of progressions and (2) relationality, through which vitality/objectivity (in questioning the bodily integration of puppet and puppeteer) is also addressed. Subsequently, I examine the seventh novel as a possible counterexample, which conversely turns out to support the initial analysis, and consider genre influences on these narrative metaphorical constructions of puppets, which provides additional perspectives on puppets in relation to both the grotesque and space.

## Control

Table 2.2 presents a more detailed breakdown of the different ways in which control is understood, framed and used across the works analysed. Notably, and in contrast to the works analysed in Chapter 1, none of these works makes explicit mention of any specific type of puppet but they focus instead on the idea of control as being central to what a puppet is. While they all established a relationship of dependency to sustain the puppet master's ability to control their puppet, a distinction can be drawn between those works that emphasised the function of control as making other people act in the way that the puppeteer does want, which I characterise here as a mediating form of control in which the puppet does the puppet master's work for them, and as making other people act in a way that goes against what the puppets

*Table 2.2* Constructions of control

| Category | Theme | Variation | Detail |
|---|---|---|---|
| Definitions | Influence | Dependency/power | Pulling strings, being ridden |
| | | Making others do as you want: speech, action (mediating) | Like a writer, director, puppet master |
| | | Making others do what they do not want to (antithetical) | Using motivations against me |
| | Monstrous | Suspense | Shadows |
| | | | Darkness |
| | | | Fear |
| | | Evil | Devil |
| | | | Spirit possession |
| | | | Eating souls |
| | | Ugly | Deformed mask |
| | | Parasitic | Hag-ridden |
| Means | Dimension | Emotional | Affection |
| | | Social | Friendship |
| | | Practical | Work-oriented |
| | | Physical | Threat/harm |
| | Knowledge | Of the puppet or the master? | Partial – a double vision |
| | C/overtness | Deception | Outright lies |
| | | | False belief |
| | | Transparency | Blatant admissions |
| | | | Outright demands |
| | Valence | Nice | Pay and perks |
| | | | Praise and affection |
| | | | Support |
| | | Nasty | Cruelty and criticism |
| | | | Threat/harm |
| Purpose | Relational | Intimate | Seeking romance |
| | | Paternalistic | Caring/nurturing |
| | Supremacy | Domination | Relative to the social |
| | | Immortality | Of the individual |
| | Retribution | Revenge | Personal vengeance |
| | | Justice | Socially situated |
| Extent | Connection | Direct: control | One-to-one relation |
| | | Indirect: manipulation | Mediated through others/events |
| | Scale | Individual | A specific character or occasion |
| | | Social | A generalised feature |
| | Degree | Graduation | All do it to some degree |
| | | | A comment on society |
| | | | Psychologised as a trait |
| | | | Socialised as effect |
| | | Progression | Escalation of: |
| | | | • Scope (number) |
| | | | • Severity (act) |
| | | | • Force (threat) |

themselves want to do, which I term here antithetical control. With mediating control, the puppetised character might or might not feel strongly against undertaking the action they are obliged to undertake (they might simply not otherwise have done it), but with antithetical control, the power exerted must be sufficient to overcome personal standards of morality and taste to compel the puppet to take actions that are abhorrent to them. Alongside these sit monstrous notions of control, usually in the context of horror or science fiction, although fear features strongly beyond these genres and darkness of the human spirit equally features in murder mysteries that do not draw on fantasy or the supernatural.

The means by which control was established and maintained varied among these novels, in terms of dimension, knowledge, covertness and valence. With regard to dimensions, the manipulative engineering of romantic or sexual relations or, in a softer form, the manipulation of social circles and relations could be considered to constitute affective dimensions, while the orchestration of work environments and scenarios to bolster a career and the use of physical threats and harm could be characterised as operational dimensions of control. Knowledge and covertness can be considered either independently or in combination, as the degree to which a puppet master is transparent about their actions and motivation will influence whether and to what extent the puppet is aware of their predicament. The valence of the strategy of control used is also related to these other aspects as, for example, a puppet master might lavish praise and affection on their puppet in an effort to engineer an intimate relationship with them in a deceptive strategy of exploitation (*Osborne*), while they might equally change their strategy from nice to nasty once their deception has been detected and they resort to direct threats and physical harm (*Redmond; Osborne*).

The purpose of this control is equally varied, characterised by three orienting themes. Relational purposes include seeking relationships of varying degrees of intimacy from friendship to sexual intercourse, and paternalistic agendas of caring for, nurturing or helping the puppet in some way, often by resolving social, familial or work dilemmas. Both can either be genuine or used as masks of deception to hide the puppet master's real self-serving agenda, as in the feigning of affection to secure sexual favours (*Osborne*). A quest for supremacy on the part of the puppet master is a second purposive theme, which also takes two forms: Social domination, for example, in securing higher status or reputation (*Redmond*), and – less commonly – immortality, which drew on the supernatural in a tale of spirit possession (*Lewis-Thompson*). Finally, retribution features strongly in these works, as protagonists seek to avenge a previous wrong done either to themselves or to somebody dear to them, whether sexual abuse, family breakdown, physical injury or incarceration. However, the novels take something of an ambivalent view on the relation between justice and revenge, with some works fluctuating between the two, both drawing a distinction between them and presenting them as equivalent and interchangeable. One protagonist deliberates on this very matter: 'It's justice. Or revenge, however you want to slice it' (*Brown and*

*DeFelice*, p88), while later the same issue crops up again: '*Revenge*. He knew the emotion well. Though for him it was more a question of justice. Justice and revenge.' (*Brown and DeFelice*, p393, italics in original). For current purposes, I associate justice more with some socially sanctioned standards of 'right' behaviour and see revenge as a more personally stimulated retaliation that need not be anchored in social standards of rightness. It is, in this reading, entirely possible for justice and revenge to be co-deliverable, but also entirely possible for them to be distinct. These different purposes are not necessarily mutually exclusive, nor are they distinct from the means of control, as – for example – social domination might be sought through work-oriented practical means in order to secure vengeance for a perceived wrong.

Under the category of extent, control can be exerted either directly in a one-to-one relationship with the puppet, which we might consider control proper, or it can be exerted in a more mediated fashion, through the orchestration of other people, events and contexts, which might be better-termed manipulation. Control can also be seen to be exerted at either an individual level, where one key character seeks to control one or more puppets (*Redmond*; *Lewis-Thompson*) or a social/collective level, both within the narrative overall where several people are seeking to control several other people (*Redmond*; *Craven*) or where society itself is being progressively controlled (*Heinlein*). It is possible to identify a graduated sense of control within and across these texts, whereby everyone seeks and secures control in certain circumstances within the norms of that society but within which society there are also extreme examples of criminal or pathological tendencies towards controlling behaviour (*Redmond*; *Lewis-Thompson*). This graduated understanding of control is depicted schematically in Figure 2.1 which draws a distinction at the extreme end of the spectrum between individual and social levels: At the individual level, extreme control is psychologised as pathological and at the social level it is politicised as ideological. The middle section in this diagram encapsulates the ways in which the characters in the novels are framed. The puppetised characters are commonly caught in a network of interpersonally problematic situations as a result of being subject to someone else's control.

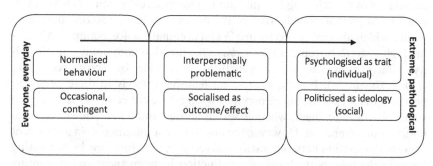

*Figure 2.1* Social graduation of control.

Source: Author.

For example, they might be uncertain how to handle a work dilemma (*Redmond*), unable to speak out about abuse due to family dynamics (*Osborne*), or unwittingly implicated in criminal acts (*Craven*; *Lewis-Thompson*; *Redmond*). The puppet masters, on the other hand, are often portrayed as exhibiting a pathological need for control due to their own previous experiences, socialised into their position as the puppet master by their own interpersonally problematic past as somebody else's puppet (*Redmond*; *Osborne*; *Craven*).

In this context, being controlled breeds a need for control, while what counts as extreme or pathological control on the part of the individual is contingent upon the behavioural norms of the society concerned. Thinking about how these two axes – individual/social and everyday/extreme – interact, we might generate a diagram like that in Figure 2.2 which indicates how an ideology of totalitarianism would recalibrate behavioural norms, exaggerating the pervasiveness and tightness of control, thereby reproducing the ideology in the everyday practice of society. Similarly, at the level of the individual, an excessive need for and exercise of control over others could generate a psychopathological diagnosis, which not only reproduces this use of control as a way of life beyond the everyday behavioural norms of society but also formalises it.

These novels, then, shine a spotlight on the susceptibility of every individual both to seek and succumb to control as well as the variety of forms that it can take, its pervasiveness, reproducibility and social specificity. They create worlds which offer biting commentary on social life and – paradoxically – both normalise and spectacularise control, highlighting the centrality of

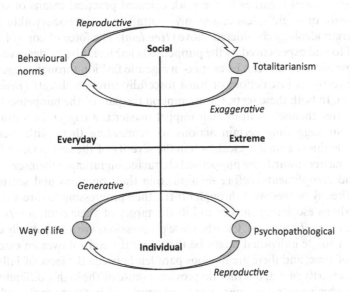

*Figure 2.2* Axes of control.

Source: Author.

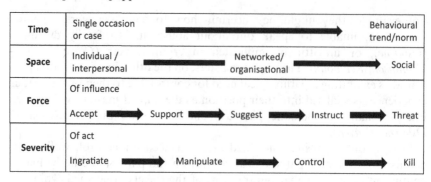

*Figure 2.3* Progression in control.
Source: Author.

contradictions to this naïve understanding of a puppet. Reflecting both the grotesqueness of puppets in being constituted through contradictions, and the potential within the grotesque for revealing deep truths or critiquing society, the latter is achieved through establishing a sense of progression in control, which itself takes several forms, as indicated in Figure 2.3.

Thinking about these progressions individually first, in terms of the temporalities that feature in these books, control can relate to a single instance or case as a reaction to a specific event or trigger or – and consistent with the psychopathological situation above – to a long-term, habituated way of relating to others. Spatially, or in scalar terms, the target or medium of control might be a single individual (*Lewis-Thompson*) but might also take networked or organisational forms as in the work-oriented practical means of control (*Redmond*), or might – less commonly – entail an entire society akin to the totalitarian ideology considered above (*Heinlein*). The force of control that is applied to and experienced by the puppetised character also evidences a sense of progression, as this can be exerted in a gentle fashion through suggestion, more directly as instruction, or most forcefully through threats (*Redmond*; *Osborne*). In both these texts, the attempt on the part of the puppetised character to free themselves from their puppet master is a trigger for a dramatic shift from suggesting certain actions to demanding them with menaces. Alongside this is a similar escalation in the severity of the action taken by the puppet master towards the puppetised character, ingratiating themselves with gifts and compliments, before manipulating their opinions and sentiments either directly or mediated through others, then progressing to direct control and perhaps escalating further to kill the target of their control (*Osborne*; *Craven*; *Lewis-Thompson*). Clearly, these progressions are not mutually exclusive, as a single individual might be the target for control over an extended period of time, and there are obvious parallels between the force of influence and the severity of act, but they do provide a sense of the highly differentiated ways in which control is constructed and employed in these novels and reinforce the critical commentary on society that is articulated within them.

## Relationality

The key characteristic of the relationship between the puppet and the puppet master in these works is evidently control, which is often established through nurturing a sense of dependency on the part of the puppetised character, but there are other ways in which we can understand how this control is established and how the puppet masters in these works forge connections between themselves and their victims, paving the way for their progressive manipulation. Despite a clear contradiction being drawn between the power of the puppet master and the powerlessness of the puppet, a range of similarities are established between the two characters, whether in terms of the locality where they grew up, their life history – for example, being in the care system – or appearance. Supplementing these similarities is the nurturing of a sense of either complementarity, in which the puppet masters position themselves as filling a void or meeting a need for their puppet (e.g., love, security, a father figure), or familiarity, whereby they establish themselves as knowing their puppets better than they know themselves or as being a fundamental feature of their social circle. At an interpersonal level, the puppet masters carve a place and vital role for themselves in the life and subjectivity of their puppet. This, in turn, is scaffolded by socially mediated and personality-focused strategies. At the social level, the puppet masters create the impression that their mode of control is normal behaviour, while their manipulation of circumstances and people surrounding their puppet both constrains their opportunities for action and creates an indebtedness on the part of the puppet towards their master. Meanwhile, the familiarity that the puppet master has with their puppet enables them to use their fears, vulnerabilities and dreams to motivate certain behaviours.

Consequently, the generation of a sense of dependency is contingent upon diverse strategies that allow and encourage the identification of the puppetised character with the puppet master, wherein the puppet master effectively makes the target of their control whole again in some way, while using their own self-identification with their puppet to identify aspects of their personality that can be exploited and using their social setting both as a mechanism of normalisation and a straightjacket of inescapability. There is, then, a strong sense of complementarity in the relationship between puppet and puppet master to juxtapose the contradictions upon which the puppet itself is constructed, and this relationship is both embedded within and scaffolded by a wider social network.

Thinking about these strategies in spatial terms, we might consider the identification of similarities between puppet and puppet master as establishing a parallelism between the two characters, while the complementarity of making the puppetised character whole again is spatial in the sense of filling a gap, either personal or social. Familiarity might be best described in terms of enmeshing the puppetised character within a reconfigured social network. At the social level, the puppet master's actions can be spatialised as being around the puppetised character, while in contrast the personality aspect is spatialised within the puppetised character. In essence, then, the puppet master shifts from being a discrete entity to being part of a bigger whole with their puppet and operates in ways that transform the external context of their puppet and/or their internal psychological tendencies, to become the indispensable glue that holds the puppet in place. The establishment and execution of control, then, can be understood as being achieved through three spatialising manoeuvres: Incorporation (becoming one with the puppet, making whole), situation (reconfiguring the social context) and disposition (moulding motivations), which collectively render the reconfigured situation inescapable by altering who they are, their opportunities for action and their referential benchmark.

This, though, points towards a different understanding of the grotesque, where the grotesque is the combination of incongruous elements, as rendering congruous is a key strategy employed by the puppet master to subjugate their victim. This can take diverse forms, such as establishing familiarity or complementarity and it need not entail the merging of bodily forms but can be subjectively enacted (e.g., either psychologically or socially), while there is also not necessarily any deformation or exaggeration of either bodily or subjective elements as the moulding of motivations might simply channel existing dispositions rather than modifying those dispositions. On a superficial reading, this might suggest that the grotesque is not therefore applicable, but – as will be argued – it is more that we need to refine our understanding of the grotesque as it applies to explicitly subjective or psychological territory.

Alongside the manoeuvrings identified above (e.g., parallelism, familiarity), there are also numerous ways in which the attitude or orientation of the puppet master to their puppet can be characterised. Underlying these orientations are perceptions of the puppet as either a creation or possession of the puppet master, forming an intentional project of construction (*Brown and DeFelice*), father–son or father-figure-type relationships (*Redmond*), an attitude of possessiveness on the part of the puppet master (*Redmond; Osborne*), or the direct possession of a puppet's body by either a spirit (*Lewis-Thompson*) or an alien species (*Heinlein*). In some cases, this 'ownership' is articulated as an effort to help or improve the puppet in some way, perhaps educating them in the ways of the world, building their career or enhancing their personal capabilities (*Redmond; Osborne*) that both infantilises and patronises the puppetised character, consistent with puppet constructions identified in Chapter 1. In other cases, the puppetised character is exploited, either for the skills and knowledge that they possess (*Craven*), for company or sexual gratification (*Osborne*) in a commodification of the puppetised character. Alternatively, this might be for the opportunity to live vicariously through the puppet (*Redmond; Lewis-Thompson*), or to exact revenge (*Osborne; Craven*), turning the puppet into a means to other ends. In yet other examples, the orientation is one of domination, reducing the capabilities of their puppet, taking away people and items that they value and acting in derogatory ways towards them to demolish their self-esteem in a trivialising process that destroys their capacity for self-determination (*Heinlein; Redmond; Osborne*).

In each case, the puppetised character is dehumanised as something that is either owned or constructed, and whose role is either to be shaped, exploited or dominated. These three orientations, then, can be conceptualised spatially in specific ways, two of which reinforce the contradictory qualities at the heart of both the grotesque and puppets, not only in the sense that they seem what they are not but also in the sense that what they say might mean the opposite. An orientation of facilitation or shaping serves as a mask for a pejorative attitude that focuses on perceived deficiencies and incapacities rather than the capabilities of the target of control: It claims to build them up while in reality it sustains their dependence on the puppet master and diminishes their potential by specifying for them where their limitations lie. This illusory space of upward mobility delivers the opposite of what it claims to deliver in a *deflection* of attitude and intent. An orientation of exploitation is not illusory in the same way, although deception might still be employed by the puppet master to get what they want; here there is a direct correspondence between the perceptions of the puppetised character by their puppet master and the puppet master's intentions and actions towards their puppet. There is no indication of trajectory here, so this is a topological or flat space of extractive resource, which – in directly corresponding to the desires of the puppet master – is a *reflection* of attitude and intent. An orientation of domination diminishes and denies any value in the puppetised character in a blatant derogation of the target of control, which inverts the duplicitous deflection of facilitation. This is an actual space of decline, imposing a sense of dejection and worthlessness onto their puppet in a self-fulfilling act that sustains the puppet's dependence upon the puppet master, which – in denying any worth in the puppetised character – is a *rejection* of attitude and intent.

Consequently, these orientations and manoeuvres on the part of the puppet master give rise to an array of puppet spatialities and spatialising processes. These constitute diverse instantiations of the grotesque that do not themselves involve bodily merging yet still consist of a two-body image (Bakhtin 1984) between the puppet and their master in which the gap between the two bodies/characters is psychologically, socially and affectively bridged. While not evident in every example considered in this chapter, there is a clear and overarching sense that the subjectivity, identity or character of the victim or puppetised character is modified in some way: They are made to feel incomplete (the puppet master portrays themselves as filling a gap), they are made to feel deficient (whether deflecting, reflecting or rejecting any specific attitude from the puppet master to their puppet) and they are manipulated into dependency upon the puppet master (whether in a work or social context). In some cases, there is direct bodily merging, but this is confined to specific genres (e.g., science fiction, the supernatural), yet the characters of puppet and puppet master are brought into alignment by the puppet master in such a way that the character of their victim is contorted in some way and to some degree. This might be evident in how they self-identify, how they relate to others, or their motivations or behaviours.

This chapter, then, while in many ways consistent with the findings of the previous chapter, also provides a more detailed understanding of the diversity and sophistication of puppet narratives and the spatialities that they generate, placing the emphasis much more firmly on the psychological than the bodily aspect of the grotesque in puppetisation. This suggests that a supplementary and complementary term to the 'body grotesque' might be helpful to accommodate circumstances in which non-bodily (or not explicitly bodily) elements of two entities are brought together in a grotesque fashion. To this end, I tease out from the analysis in this chapter the notion of the 'character contorted': While possibly but not necessarily accompanied by bodily merging, the puppetised characters analysed in this chapter frequently have their character modified or at least manipulated by the strategies and orientations of their tormentor, which serve to sustain their torment. While the body grotesque and the character contorted are not necessarily mutually exclusive, they are also not necessarily concurrent, and – in addition – the character contorted, as established in this chapter, does not necessarily require the hyperbolic or excessive quality associated with the body grotesque (Bakhtin 1984; Li 2009; Duggan 2016), although this is not precluded. Indeed, most of the puppet masters explored in this chapter maintained their control over their victims precisely because their attitudes/orientations to their targets and the strategies they employed to control them were subtle and predicated on rendering congruent the two bodies/characters concerned, making the grotesquery less obvious and constructed as much on complementarity as on contradiction. Seemingly, then, there is a need both for a more refined conceptualisation of the grotesque that goes beyond the obvious excess of merging incongruous entities to accommodate the more subtle contortion of entities through strategies that render their constitutive parts congruent rather than contradictory, and for a supplementary term to sit alongside the body grotesque to accommodate more explicit subjective, psychological and social reformulations of the two-body image. The character contorted – as introduced and empirically grounded in this chapter – provides such accommodation.

In addition, the orientations, manoeuvres and strategies discussed above provide further insights into popular cultural constructions of puppets by attending to the puppet master as much as to the puppet and in drawing equivalences between the individual and the collective, while also establishing a graduation from the individual to the collective as a means of delivering both an interrogation of the human and social commentary. There is, though, one final narrative metaphorical use of puppets worth examining, which while displaying many of the same themes as the six works examined in this chapter, employs them in different ways that suggest the value of this work as a potential counterexample to test the robustness of the previous analyses. Through an analysis of this final work – Danielle Ramsay's (2016) *The Puppet Maker* – it will be argued that the discrepancies in how this work constructs the puppet do not so much challenge the analyses presented in these first two chapters as they suggest that the titular emphasis on the puppet in this novel is misplaced.

## A compelling counterexample?

Ramsay's (2016) *The Puppet Maker* is especially informative to this analysis as despite its titular emphasis on 'puppet' and despite some evidence of core contradictions that might typically be associated with puppets (and the grotesque), this work seemingly disconfirms the detailed findings of the foregoing discussion. For example, the theme of control is clear in this novel, and while there is some evidence of one character (a police officer) being controlled through the manipulation of their social context by another, the primary locus of control is the body of each young female victim, seemingly switching the emphasis back to the bodily aspect that dominated in Chapter 1. As will be shown, however, more aspects of this seventh novel reinforce than unsettle the emergent findings, such as mobility, sociality, relationality and faciality.

In this seventh novel (*Ramsay*), the initial relationship identified between control and mobility is reversed, as the so-called puppets are immobilised by being strapped to a chair in which they are subjected to horrific 'medical' procedures to 'cleanse' them through lobotomy. In all other works analysed the default position was that the puppetised character was made to move in a jerky fashion, not prevented from moving, so in this way, this work is inconsistent with the others. Similarly, the socially mediated nature of control established by puppet masters inserting themselves into and manipulating the social world of the puppetised character is absent in relation to the female victims in this work. Instead, the victims are vulnerable to being puppetised because they are already marginal to society and their extraction from society through kidnapping enables their 'puppetisation'. Thus, although control features in this novel, consistent with both analyses presented in Part 1, and both mobility and a bodily rather than psycho-social emphasis are evident, consistent with the analysis in Chapter 1, the relationship between control and mobility is inconsistent with that outlined in Chapter 1 and the prioritisation of control through social evacuation rather than manipulation is inconsistent with the analysis in Chapter 2.

In addition, the most significant relationship in this work is not that between the 'puppet' and its master, but between the master and his family, as the puppet master is re-enacting the past of his own family by inflicting past familial acts on victims beyond that family circle. The father's medical expertise is re-enacted, the mother's chair is the site of torture and the sister's doll is the image in which the victims are recrafted. In recrafting the victims in the image of his sister's doll, the 'puppet master' is not treating these young women as a means to achieve a greater end, as in the other six works, but as a target for bodily reshaping, and there is no integration – bodily or psychological – between the 'puppet' and its maker. There is no meaningful relationship between the puppet maker and his victims, contradicting the relationality in the other six narrative metaphorical engagements with puppets. Further, the titular emphasis on making rather than mastering the supposed puppets places the emphasis on the female victims (and the doll) rather than the police officer, who is more appropriately considered to be being mastered (manipulated) like a puppet. Moreover, in terms of visuality, these victims are described as having perfect faces, a far cry from the deep lines and gaping jaws that characterised the puppet's appearance in Chapter 1, while in terms of mobility, these victims are described as being dressed and positioned identically, reinforcing the lack of mobility rather than the archetypal mobilities of puppets identified in Chapter 1. In the everyday or naïve understanding of puppets, puppets are not arranged but are operated; they are not prevented from moving but are made to move in highly stylised ways; and they are not positioned but they slump when they are not moving (e.g., when their strings are cut).

It is in relation to contradictions that this work most closely aligns with the themes identified as constitutive of puppets in this chapter, as the puppetised victims are described as appearing both lifelike and unnatural, as being unique yet identical and as crossing the distinction between doll and girl, in a classic example of the contradictoriness of the grotesque (Kayser 1963; Harpham 1976, 2006; Danow 1995; Connelly 2012). However, the role played by the sister's doll in this narrative – in providing the desired appearance to be achieved with each victim – clearly establishes that the protagonist is making dolls rather than puppets. This, allied with the discrepancies in how the victims in this work are constructed compared to those in the other works analysed, serves to undermine the appropriateness of the metaphorical use of the puppet as the narrative focus for this work. Consequently, despite the repeated assertions in this novel that the press-given name 'The Puppet Maker' is apt and fits perfectly, this analysis suggests that this is not the case and that 'The Doll Maker' would be more appropriate as a title. The secondary storyline of the manipulation of the contacts and circumstances of the police officer is more consistent with the understanding of the puppet constructed through the analysis in Chapter 2, yet it is not this police officer who is being 'made' by the protagonist but the female victims that are being made into dolls. This seventh narrative metaphorical text, then, despite seeming to serve as a counterexample, conversely confirms the earlier analytical findings. Its greater affiliation with dolls than puppets strengthens the case presented here for the key characteristics of the naïve, everyday understanding of puppets as evidenced in their passing and narrative metaphorical use for fictional purposes.

## Genre influences

So far, I have focused on these novels as a collection to explore a generalised understanding of puppets, but if we remove the genre influences of science fiction and the supernatural, the psycho-social emphasis in these constructions of puppets becomes even more apparent as it is only in these genres that bodily merging occurs, either through (parasitic) attachment or (spirit) possession. In purely human settings, puppets are constructed either through analogy or obsession, which I discuss here in relation to three themes: How it works, the effect it has and the spatiality that emerges.

Analogical approaches feature protagonists who control other people as if they were puppets, with an emphasis on social and psychological means of control that are exercised primarily in a mediated form through settings and social networks, with the psychological aspect concentrated on the puppet master's knowledge of the puppetised character's psyche, enabling them to manipulate (or contort) that psyche. The obsession approach is similar, in that it is primarily psycho-social in nature and primarily mediated through others and circumstances, but in this instance, the psychological emphasis is on the puppet master at least as much as their puppet. This approach involves the puppet master projecting something of themselves onto their victim beyond simple similarity, for example, as a younger version of themselves through whom they seek a second chance at life (again, potentially – but not necessarily – a character contorted). By contrast, with both attachment and possession, the character might – or might not – be contorted and this might – or might not – be accompanied by a 'grotesque-ing' of the body.

While all approaches involve the eviction of subjectivity identified in Chapter 1, this takes different forms in each case. With attachment, the human hosts cease to think for themselves as the directness of cognitive control erases any capacity for self-determination, while the possession approach entails one body accommodating two personae, with competition and fluctuation between the two. Significantly, the analogous approach does not involve any impact on the cognitive capacities of the puppet, but contexts and events are orchestrated in such a way that the puppet's actions can be manipulated by exploiting rather than modifying their motivations and thinking style. Here, it is not necessarily the character that is being contorted, but the social/familial/professional/affective relations that are being contorted, which thereby enables the puppet master to control their victim seemingly without grotesque-ing either their body or their character. By contrast, the obsession approach incorporates a combination of the adaptation and overriding of thought processes. In this case, thoughts and feelings are initially adapted as the puppets seek the approval of their master, so these puppetised characters continue to think for themselves and make their own decisions but in skewed ways, yet once the puppets seek to reassert their individuality and independence, the puppet master crudely overrides the puppet's cognitive and affective processes through force in a case of simple, knowing submission.

Finally, these different treatments of the puppet evoke different spatialities in terms of how the puppet and its master relate to each other. In the analogous approach, the puppet and puppet master remain entirely separate entities, even though the puppet is psychologically and emotionally yoked to its master. It is the web of social and contextual connections within which they are both enmeshed that is key to the exercise of power, establishing a networked but distanced spatiality of (potentially) a character contorted. With the attachment approach, the puppet master is literally a bodily attachment, forming a cerebral conduit through which control is exercised in a spatiality of adjacency and integration. With the possession approach, the puppet master resides entirely within the puppet as an internal secondary personality that forms a semi-cooperative conscience as both puppet and master share certain ambitions. The obsession approach reverses this, however, as in this case, there is one psyche but two bodies: The overriding of the puppet's thought processes effectively means that their respective consciousnesses coincide, with two bodily forms acting in accordance with one dominant personality.

While the grotesque is clearly evident in the varied mergings of the bodily and the psychological, the psychological, social and affective emphasis evident in these works spotlights the more-than-bodily aspect of the grotesque (Duggan 2016) which – in this context – seems considerably under-theorised compared to the body grotesque, and the character contorted is proposed as a supplementary term to the body grotesque to accommodate such emphases. Moreover, these puppet constructions draw out two alternative instantiations of the grotesque. On the one hand, there are two examples of a more distributed understanding of the two-body image central to the grotesque (Bakhtin 1984) as the connections between the two bodies are distributed through a network of social relations (e.g., yoking two characters into one unifying consciousness and/or contorting those relations to manipulate the puppetised character). On the other hand, we can distinguish between the grotesquery of the puppet as constructed on contradictions and the grotesquery of the puppet–puppet master duality as constructed on both contradictions and complementarities. This chapter then, not only unpacks additional and different constructions of the puppet and specifies an increasing number and diversity of puppet spatialities compared to the previous chapter, but also advances our conceptual lexicon by introducing the character contorted, which can accommodate explicitly psycho-social articulations, more distributed forms and alternative 'rendering congruous' strategies of the grotesque.

## Conclusion

This chapter has developed the introductory analysis of everyday understandings of puppets, as portrayed in their metaphorical use in adult popular fiction. The consistency among the themes identified in each of the works led to the organisation of the discussion around control and relationality before consideration was given to a possible counterexample and the role of genre in determining the construction of the puppet in these narrative works.

Control was examined in terms of its definitions, means, purpose and extent, giving rise to a more detailed classification of this crucial aspect of puppet narratives and fleshing out more nuance in the relationship between the puppet and the puppet master – for example, in relation to emotional valence and the distinction between justice and revenge – than emerged from the analysis in Chapter 1. Picking up on disciplinary themes of structure and agency, scale, the everyday, and power and resistance, this discussion led to an appreciation of the social commentary embedded within this set of works and the scalar nature of control in relation to time, space, force and severity, which in combination evidence the potential of the grotesque to deliver social critique through puppet constructions, both of which are revisited in the Conclusion to Part 1. Reflecting on the role of control in social life and the continuum from individual to societal tendencies to control or be controlled, the ideology of a society sets the behavioural norms for interpersonal relations and psychopathological instances of control are identified against those norms, both reinforcing and challenging those social norms.

Subsequently, the ways in which a relationship is established between a puppet and its master to enable that control to be exerted was explored through the consideration of diverse manoeuvres or strategies employed by the puppet master and different orientations adopted by the puppet master to their puppet/s. Discussion of the strategies by which relations of control were established led to the identification of three spatial manoeuvres or spatialising processes by which the puppet master positioned themselves with respect to the victim: *Incorporation*, an entering into the life and world of the puppet to become a vital part of their self-identity; *situation*, an intervention into the puppet's familial, social and occupational context, and *disposition*, an internalisation of control through the manipulation of the puppet's cognitive and affective processes. In each case, the puppet master crosses a border or threshold to exert their influence over their puppet: Into the life of the puppet, into the social circle of the puppet or into the psyche of the puppet, but in none of these is there a conjoining or transposing of bodily materialities, in an alternative configuration of the grotesque in which the life, sociality and psyche of the puppetised character are rendered grotesque, rather than their bodily form, which is conceptualised here as a character contorted.

Consideration of the different attitudes or orientations that a puppet master takes towards its victim led to the identification of three spaces of varying intentional duplicity: *Deflection*, where the actual intention is masked by well-meaning rhetoric; *reflection*, in which there is a direct correspondence between the real intentions and the indications of those intentions, and *rejection*, where the intention is attained through the denial of any intention at all. While there is evidence within this discussion of the topographical nature of the body grotesque (Bakhtin 1984) in that – for example – there is illusory upward mobility alongside denigrating degradation in deflection, the second (reflection) is topological rather than topographical as there is no normative or affective inflection to the rhetoric of the puppet master, just straightforward intent. These spatialities both modify the topographical emphasis of the body grotesque and enact and reflect the contrariness of puppets (and the contradictoriness of the grotesque) as – with the exception of reflection – the attitude towards the puppet projected by the puppet master is not what it appears.

The analysis of the seventh novel initially appeared to conflict with the findings of the first two chapters as it constructed the control, mobility and relationality of the supposedly puppetised characters in ways that contradicted those features of the naïve understanding of the puppet previously identified. However, when read in the context of the dominance of the sister's doll in the primary plot, these seeming contradictions cast new light on this novel, suggesting that the title is misplaced, being more appropriate to the secondary plot than the main narrative. Ultimately, through this exploration of a potential counterexample, the findings established in the first two chapters of this volume are reinforced rather than challenged by this seventh novel.

Finally, discussion of genre influences through a typology of puppet narratives found in these works generated four different approaches and four further spatialities of puppet–puppet master integration, each of which relates in its own way to the strategies used in the construction of the character contorted: *Analogous* narratives in which the puppet and its master are separate but yoked in a connective web of relationality; *attachment* narratives in which the puppet master is adjacent to and partially integrated with the puppet (concurrently forming a body grotesque); *possession* narratives in which an additional persona is internally accommodated within the puppet's body in a semi-cooperative consciousness; and *obsession* narratives in which one psyche is extended over two bodies in a coincidence of consciousness as the puppet master takes over. Reflecting yet more diverse constructions of the grotesque, these bring us full circle to the acknowledgement of both the bodily and psychological excess that characterise the grotesque (Duggan 2016), which take specific forms in the constructions of puppets, prompting the proposal of the character contorted to supplement the body grotesque.

The significance of this last set of spatialities lies in two additional perspectives that it brings to bear, which reinforce aspects of the analysis in the previous chapter. The first relates to the hollowing out of individuality identified but under-developed in the passing metaphorical uses of puppets, as the narrative metaphorical uses indicate diverse forms that this can take, confirming this as an important consideration in disciplinary contexts. In some cases, individuality is adapted to align with the ambitions of the puppet master, in others it either lives alongside or competes with the puppet master's character, and in yet others, it is simply overridden, bringing a more nuanced sense of how this hollowing out can take effect. The second relates to the importance of engaging with the respective materialities of puppet and human bodies, which is conspicuously absent from the analysis in this chapter. While there is some evidence of bodily merging between a puppet and its master, this featured in specific genres rather than explicitly engaging with this aspect of puppets. Within entirely human plots, the emphasis is very much on psycho-social modes of control, although physical threats and injury also feature, such that bodily emphasis and material transposition are omitted from consideration. The spatially mediated nature of this psycho-social control unites ideas of biopolitics and geopolitics and applies them at the interpersonal level, providing an alternative appreciation of these disciplinary concerns, which is developed further in the Conclusion to Part 1.

Nonetheless, the analysis in this chapter establishes that narrative metaphorical uses of puppets employ contrary aspects of puppets and unsettle a range of borders between puppet and puppet master. While this is consistent in broad terms with scholarship on the grotesque in terms of contradictions and borders, especially in the centrality of a two-body image, it also injects different perspectives on the grotesque. For example, it does not always involve a combined material bodily form, the orientations of/between those bodies/characters can be as much topological as topographical, and the grotesque – especially in the context of the character contorted – does not necessarily entail excessive hyperbole but can be achieved through subtle 'rendering congruous'. This analysis, then, supplements those laid out in Chapter 1, highlighting how these varied strategies initiate spatialising processes on the part of the puppet master that create multiple, diverse bodies grotesque and characters contorted. Furthermore, it highlights how puppet spatialities can deliver both psychological insight and social commentary and contribute to the conceptual refinement of the grotesque.

## References

Bakhtin, M. (1984) *Rabelais and his world*. Indiana University Press: Bloomington, IN.

Connelly, F.S. (2012) *The grotesque in Western art and culture: The image at play*. Cambridge University Press: New York.

Danow, D. (1995) *The spirit of carnival: Magical realism and the grotesque*. The University Press of Kentucky: Lexington, KY.

Duggan, R. (2016) *The grotesque in contemporary British fiction*. Manchester University Press: Manchester.

Harpham, G. (1976) The grotesque: First principles. *The Journal of Aesthetics and Art Criticism*, 34(4): 461–468.

Harpham, G. (2006) *On the grotesque: Strategies of contradiction in art and literature*. The Davies Group Publishers: Aurora, CO.

Kayser, W. (1963) *The grotesque in art and literature*. Indiana University Press: Bloomington, IN.

Li, M.O. (2009) *Ambiguous bodies: Reading the grotesque in Japanese Setsuwa Tales*. Stanford University Press: Stanford, CA.

### *Literary works analysed*

Brown, D. & DeFelice, J. (2016) *Puppet master*. William Morrow: New York.

Craven, M.W. (2019) *The puppet show*. Constable: London.

Heinlein, R.A. (1951/2010) *The puppet masters*. Baen Books: New York.

Lewis-Thompson, J. (2019) *The puppet master*. Beacon Hill Publishing, no place of publication listed.

Osborne, A. (2017) *The puppet master*. Bloodhound Books, no place of publication listed.

Ramsay, D. (2016) *The puppet maker*. Mulholland Books: London.

Redmond, P. (2000) *The puppet show*. Hodder and Stoughton: London.

# Conclusion to Part 1

The two chapters that constitute Part 1 have both looked at metaphorical uses of puppets – primarily in adult fiction – to unearth an everyday understanding of the puppet. The first chapter examined passing metaphorical references to puppets that were used to convey aspects of a character or the circumstances in which they find themselves, while the second explored more substantive uses of the puppet as an orienting device for the whole narrative. Each chapter revealed similarities in terms of the control-based nature of the relationship established between the puppetised character and its 'master', the muddling of their bodily or subjective borders and the specific spatialities that this muddling generated. Despite these similarities, the two chapters also identified different degrees to which the sets of works analysed engaged with specific types of puppets and different emphases in the way in which the relationship of control was established. These reflected differences in how each set of works engaged with the muddling of bodily and subjective borders and led to a specific and varied set of ideas about the puppet as grotesque and about the grotesque as a concept. In this conclusion to Part 1, I begin by specifying and refining the implications of Chapters 1 and 2 through a comparative analysis of their findings to provide an initial answer to the research question guiding this part: What is a puppet? Subsequently, I elaborate further on the relationship between puppets and the grotesque in a geographical context and finally lay out and collate the puppet geographies – or spaces of puppets – that emerge from Part 1.

## Comparative analysis

Each of the first two chapters revealed its own naïve understanding of the puppet, which in some ways were consistent with each other and in some ways differed. What remains to be considered, however, is what these two sets of analyses tell us in combination about puppets in popular culture and their varied geographies. The exploration of passing metaphorical references revealed a composite idea of the puppet comprising the head of a ventriloquist's dummy but the body of a marionette with exaggeratedly ugly features and erratic jerky movement, while the transposition of wooden and fleshy bodily materialities was used for graphic and affective purposes in an

DOI: 10.4324/9781003214861-6

incompatible combination of the disavowal of human characteristics (subjectivity, fleshiness) and the exaggeration of bodily and psychological brutality. The strings of the marionette are the metaphorical means by which the relation of control is depicted and the metaphorical puppetisation in these works is deemed to deindividualise the character concerned, although the nature of this deindividualisation is not thoroughly articulated, and it enhances rather than diminishes the puppet's capacity to evoke empathy.

The materiality and faciality that were central to passing metaphorical references are conspicuous by their absence in the narrative uses of puppets, with the exception of the faciality of the victims in Ramsay's *The Puppet Maker*, which contradicted the puppet's grotesque appearance identified in Chapter 1 and which was associated more with dolls than puppets. Setting this work to one side for this reason, from the first to the second analysis, the emphasis shifted from a focus on the body of the puppet with control largely considered as an undifferentiated notion, to a central concern for the specifics of control (means, purposes, degree, impacts) without direct consideration of types of puppet, their facial appearance, their bodily materiality or their movement capacities.

A more strongly psychologised and socialised construction of puppets emerged from the analysis of their narrative metaphorical uses, especially in terms of the conjoining of puppet and master, except for the science fiction and fantasy/horror novels, which are influenced by their specific genres more than the theme of the puppet. The relative lack of consideration of materiality in narrative uses precludes a more thorough exploration of such opportunities for storytelling or graphic effect, establishing a clear distinction between the passing and narrative metaphorical uses of puppets in adult fiction. Similarly, but in reverse, the emphasis on the body/materiality of the puppetised character in passing references prevents the unpacking of different forms that the hollowing out of the individual might take, even though it is identified as characteristic of that process of puppetisation. Instead, this comes through much more robustly and in diverse ways within the narrative uses of puppets.

Overall, the passing metaphorical references to puppets generate a composite and simplistic, body-focused view of the puppet as an object of control that paradoxically both humanises and objectivises the puppetised character to generate empathy in the reader, while the narrative metaphorical uses of puppets construct a differentiated, complicated and more psycho-social view of the puppet that paradoxically combines the normalisation and spectacularisation of control within both the puppetised plot and the everyday behavioural norms of society, providing critical commentary on society.

One aspect that was consistent in both opening chapters is how remote these naïve understandings of puppets are from contemporary puppetry practice and the articulated perspectives of puppetry practitioners and experts (designers, makers, performers, critics). The emphasis on puppets as small scale, wooden and the object of control all differ from the diverse sizes and materialities of puppets as currently employed and the sense of puppets as having their own agency that emerges from accounts of puppetry practice (Tillis 1996; Bicât 2007; Gross 2011; Francis 2012). There are, then, at least two distinct understandings of puppets in popular Anglophone culture already emerging – that of practitioners and that of novelists – with the latter containing considerable diversity revolving around a central concern for control and articulated through specific (re)configurations of bodily and psychological borders that are easily construed as grotesque.

### Puppets and the grotesque

As established in the Introduction, associations between puppets and the grotesque are both historic and strong, and the analyses presented in Chapters 1 and 2 confirm considerable consistency between the metaphorical references to puppets and understandings of the grotesque. There is a clear juxtaposition of the playful and the horrific, evident excesses of both a bodily and psychological nature (e.g., the ugliness of the puppet, the monstrosity of the puppet master), and an open, social and unfinished puppet body suspended in a web of manipulable social relations (Kayser 1963; Harpham 1976, 2006; Bakhtin 1984; Connelly 2012; Duggan 2016). As Figure C1.1 illustrates, these puppet constructions are also founded on a range of incompatible categories that are variously brought together, such as puppet and human, subject and object, contradiction and commonality, thereby grounding these puppet constructions firmly in the grotesque.

Beyond such consistencies, however, these analyses also cast new light on engagements with and understandings of the grotesque. While there is very definitely a two-body image (Bakhtin 1984) at play in these constructions of puppets, this is sometimes extended to three bodies (two types of puppet and

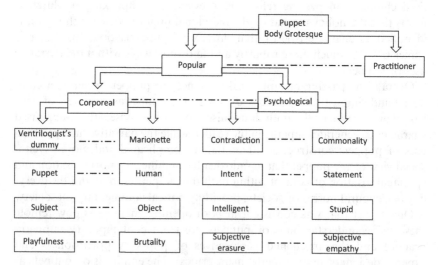

*Figure C1.1* The body grotesque of the puppet.

Source: Author.

the human form), it is sometimes not directly founded on the physicality or materiality of the body (as in psycho-social narratives), and an important aspect of the two-body image identified in each chapter is the distinction between the practitioner and popular understandings of the puppet. This latter issue is interesting as it introduces the possibility for the body grotesque not to be formulated by the merging of different bodily forms but by the bridging of two different perspectives on the body, with each perspective being sustained yet mutually incompatible (e.g., the puppet as both agentive and not agentive). The emphasis of this book, then, is very much on popular constructions of puppets rather than practitioner constructions, and the analyses in Part 1 evidence a firm distinction between constructions of the puppet that focus on the corporeal and those that give primacy to the psychological, establishing another combination of contradictory emphases.

Another significant way in which the body grotesque of the puppet – as constructed in metaphorical fictional references – differs from that of conventional understandings of the body grotesque is that the body grotesque of the puppet is at least as much topological as topographical. There is an emphasis on the face in passing metaphorical references to puppets and there are some ascending and descending trajectories identified in terms of puppet master orientations towards their victims. However, there is no countervailing emphasis on the lower half of the body and these trajectories were more about interpersonal attitudes in relation to social mobility than bodily movement, whereas in conventional understandings of the body grotesque, bodily movements are employed to indicate or potentiate social mobility at a more societal level.

Also important here is the emphasis in topological understandings of space on connections rather than distance (Ghoddousi and Page 2020), which brings us to the significance of a shift in mindset from gaps to borders between the contradictory categories employed when we think about puppets in relation to the grotesque. One important aspect of a border is that it connects the things that it divides as much as it divides them, so fits neatly within the contradictory terrain of the grotesque. On this reading, the dashed lines in Figure C1.1 between paired contradictory terms do not indicate distance or separation but contact and connection. While logical or rational gulfs remain between contradictory conceptual categories employed in the construction of puppets, those categories are topologically united as they are connected in the counterintuitive construction of the puppet.

Progressing from a notion of the grotesque as a concept without form to a form that challenges conceptualisation, the form is forged through the recrafting of the border between the conceptual categories and the rampant interpenetration of those categories among each other, such as between a ventriloquist's dummy's head, a marionette's body and a human being. In Chapter 1, this was found to be most powerful in spatialities 2 and 3 of Figure 1.1, when the human/object distinction was most confused, especially if the border is muddled at both bodily and psychological levels. In this circumstance, the contradictory connections are densest, but the incompatibility of the conceptual contradictions remains. In this case, the contradictoriness of the grotesque is not concerned with conceptual distance but with the potential of reconfiguring the border between the categories to bring them together in a contrastive unity. These incompatible conceptual categories are united by the very terms of their incompatibility. Consequently, it seems fitting to move away from notions of categorical multiplicities and gaps (Harpham 2006) and towards an appreciation of conceptual gulfs that persist between categories, but which are topologically united through creative processes of reconfiguring the border. These creative acts are not drawing borders anew to separate or distinguish between categories but are redrawing borders to unsettle that separation even while reinforcing the distinctiveness between the conceptual categories that are brought together, through interrogating them in their mutually incompatible connectedness.

The grotesque, here, is both an act of construction of the puppet and actively reconstructed through the puppet. The space between conceptual categories in the grotesque is thus reconfigured as the gap is replaced by a gulf that persists conceptually but is overcome topologically through an emphasis on contradictory connections rather than the distance of difference. This, in turn, brings us to the significance of space more generally, which incorporates puppet spatialities, spatialising processes and the spaces of puppets, bodies grotesque and borders.

## Puppet geographies

Chapters 1 and 2 each identified their own puppet spatialities that were associated with the diverse ways in which the puppets themselves were constructed. These are presented in overview in Figure C1.2 with the genre-specific

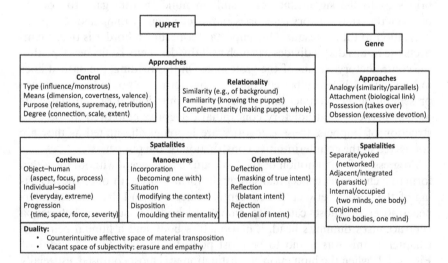

*Figure C1.2* Constructions and spatialities of the metaphorical puppet.

Source: Author.

approaches and spatialities to the right in acknowledgement that while constructing their own puppets some aspects of these constructions were clearly rooted in genre conventions rather than necessarily contributing to a more generic understanding of the puppet. The bulk of the diagram outlines the key defining features of puppet constructions as identified in these analyses – control over the puppet and the relationality between the puppet and its puppet master – and lays out the myriad ways in which these constructions can be considered spatial. These include continua, such as from the puppet as object to the puppet as human, the manoeuvres and orientations enacted by the puppet master, and the powerful duality between the objectification of the material body and the evacuation of subjectivity that evokes empathy even while erasing the individuality and humanity of the character with which we empathise.

The analysis of passing metaphorical references generated a continuum of four overlapping spatialities, resulting from different combinations of aspect (bodily/psychological), focus (form/material/subjectivity) and process (merging, transposing, evicting and extending), with the most powerful graphic and affective outcomes arising from the multiple and complex intersplicing of puppet and human features through the muddling or recrafting of puppet–human borders, both bodily and subjective, as explored in the previous section. The analysis of narrative metaphorical uses of puppets identified three different sets of spatialising processes, through which the relationship between the puppetised character and its puppet master is established:

1. Strategic or control-oriented manoeuvres of incorporation, situation and disposition.
2. Orientational processes of deflecting, reflecting and rejecting the attitude of the puppet master to the puppetised character.
3. Processes of analogy, attachment, possession and obsession identified through the typology of narrative uses of puppets across the different genres.

Similarly, the borders that are muddled in the generation of these spatialities are multiplied and diversified in Chapter 2 compared to Chapter 1, which focused on bodily and subjective borders in generating the affective power of puppet metaphors. While bodily and subjective borders continued to feature in Chapter 2, further borders were added, including social borders that are confused or reconfigured as the puppet master seeks to control their puppet through manipulating those around them and the border between reality and deception in the orientational spatialising processes. The bodily merging of puppet and puppet master also found new forms of expression in Chapter 2 through attachment and possession, although these were largely influenced by the genre of the work rather than being explicitly associated with puppets. Just as control was found in Chapter 2 to be progressive in character, so too, this analysis uncovered a progressively more sophisticated and increasingly differentiated understanding of the puppet and its varied spatialities as the engagement with the nature of the puppet forms a stronger element in the plot.

These findings also speak to broader disciplinary concerns. In the context of more-than-human and non-representational geographies' interest in decentring the subject (Simonsen 2013; Simpson 2017), the simultaneous eviction of the subject and affective focus on the subject through the puppet's status as simultaneously more-than-human and less-than-human potentially positions the puppet as a conceptual and material tool through which to further these debates (see Part 3 for further development of this potential). Similarly, while the analyses in Part 1 evidence the acknowledged role of puppets in providing a metaphor for the human condition and human society, they also hold specific disciplinary significance in doing so. On the one hand, the bidirectional scaling effect whereby the puppetised character is simultaneously expanded through its distribution across the network of control and dwarfed by that very network sits both comfortably and uneasily among contemporary concerns for more topological understandings of space that overcome distinct scales (Ghoddousi and Page 2020). It is a comfortable fit in that the simultaneous bidirectionality suggests scalar flexing rather than scalar categories, but it is an uneasy fit because this flexing reintroduces more categorical understandings of scale. On the other hand, the scaling up of relations of control from the individual to the social not only draws attention to disciplinary dualisms of structure/agency and individual/collective, but also enables the use of puppets and puppet narratives for communicating social commentary and critique. Core features of everyday understandings of puppets potentially position them well within a discipline with a historic but difficult relationship with critical scholarship and a desire to change the world for the better (see, for example, Blomley 2006, 2007, 2008), especially in the current climate of the impact agenda's formal demand for academics to generate such change (Rogers et al. 2014) and associated drives towards greater public engagement with and through scholarship (Murphy 2006; Ward 2006; Kitchin 2014).

Perhaps most informatively, though, there is an equivalent scaling down of certain concepts and theorisations usually applied at the societal level to the level of the individual and interpersonal. The spatially mediated nature of these psycho-social modes of control shines a spotlight on the operation and weaponisation of geopolitics and biopolitics at the interpersonal level. Of these, geopolitics refers to a set of practices through which geographical knowledge is employed within technologies of power as applied to territorial space (where that employment is termed geopower) (O'Tuathail 1996), including the imaginations and representations on which international political relations are predicated (Gregory et al. 2009; Reuber 2009). Similarly, biopolitics can be considered a set of practices through which biological knowledge (of the human) is employed within technologies of power as applied to the governance of a population (where that employment is termed biopower), framed as protecting that society from an internal threat and intervening in the lives of individuals to protect the interests of the collective (Foucault 2007; Anderson 2012; Philo 2012). While these two forms of power

seem quite distinct, there are considerable overlaps between them, as the geopolitical imaginaries are – in part – perpetuated to influence the attitudes and
behaviours of the populous with respect to the demonised external Other.
Equally, the biopolitical – although not territorial (Philo 2012) – is bound up
with spatialities of surveillance, such as the panopticon that secures behavioural conformity because the uncertainty as to whether surveillance is active
encourages behaviour as if it is, and the spaces through which behavioural
norms are both supported (e.g., hospitals and schools to generate healthy,
educated and therefore productive citizens) and enforced (e.g., courts and
prisons to sanction individuals who transgress behavioural norms and discipline the populous at large). Moreover, the integration of biopolitics at the
level of the individual into the more general technologies of governance at
the level of the population has also been described as establishing a geographical emphasis in biopolitics (Lehtinen 2009).

The biopolitical and the geopolitical, then, are more closely entwined
than we might commonly assume, and this is evident in the analyses in Part
1. In many instances, the puppet master secured control of their victim (biopolitics) through the manipulation of the spaces (e.g., social, professional)
they inhabited (geopolitics) and/or through the adoption of spatial and spatialised strategies (e.g., inserting themselves into the world/s of their victim)
to alter the behaviour of their victim by modifying what they consider to be
normal or desired behaviour (biopolitics). The two very much work in tandem and although often enacted through social networks, this is at a more
local level than conventional framings of geopolitics, which is explicitly
about large scale spatialities, conventionally global space (O'Tuathail 1996;
Reuber 2009). The varied strategies of control employed by the puppet masters in these fictional works and the diverse spatialities that are performatively created and reshaped through these strategies thus encourage a more
fine-grained consideration of the social and interpersonal function of geopolitics and biopolitics in combination. While not wishing to demean the
scalar specificity of geopolitics as formally conceived, the linking of society,
space and power within geopolitics can be seen in the puppet narratives here
on an interpersonal level and although geopolitics proper might be considered inappropriate in this context, the intertwining of geopolitical strategies
and the simultaneous enactment of biopolitics at the level of the collective
and the individual brings at least certain aspects of geopolitics down to the
interpersonal scale. In this context, it is fruitful to consider these puppet
narratives in terms of geobiographies and biogeopolitics (Lehtinen 2009),
where the former is attributed to the puppetised victim as the modification
of their biography (life story) through the interpersonal geopolitical strategies of the puppet master and the latter is attributed to the puppet master in
recognition of the combined biopolitical and geopolitical strategies (and the
interweaving of each through the other) through which they generate that
modification. These conceptualisations, then, encapsulate how the geopolitical and the biopolitical are deeply and mutually embedded in these puppet narratives, from the individual and interpersonal to the social and

societal, and this embeddednesss is important in facilitating the socially critical potential of puppets.

In addition to these explicit puppet spatialities, we also need to acknowledge the central importance of the spaces of borders and bodies, and the importance of borders and bodies in the creation of puppet spaces, as elaborated in the preceding section. Borders and bodies are of significance in relation to puppets not just because they are foundational to the grotesque and the grotesque is associated with puppets but because borders and bodies are implicated in constructions of puppets in particular ways, which prompts a rethinking of certain aspects of the grotesque. This has expanded our appreciation of the two-body image/space, shifted our spatial attention away from gaps and towards gulfs and modified our spatial theorising away from conceptual distance to topological connections, which are forged through creative practices of borderscaping. In the context of these many and varied intersections between puppets, spaces, politics, concepts and geography it becomes increasingly clear that – from understandings of space, through disciplinary engagements with the body and the subject, to aesthetic theory and social theoretical perspectives – puppets hold considerable generative potential for geography.

## Conclusion

Evidently, even everyday metaphorical uses of puppets in popular fiction reveal their own geographies, which are multiple, diverse and contingent upon the specifics of the narrative under consideration, immediately establishing puppets as a ripe field for cultural geographical analysis. Moreover, several points of connection between metaphorical uses of puppets and geographical themes and interests both within and beyond cultural geography have been identified including visuality and materiality; the body; mobility; subjectivity; affect; more-than-human or socio-material assemblages; scale and scaling effects; dualisms (individual/collective, structure/agency); psycho-geography (in the continuum of control from the psychological to the social); geopolitics and biopolitics; and borders.

Although the constructions of puppets and their spatialities are more sophisticated in their narrative use compared to their passing reference, the latter rather than the former makes more intriguing use of the materiality of the puppet, with implications for disciplinary engagement with posthumanism, while the former rather than the latter provides greater potential for critical social commentary, with implications for geographical work on the confluence between biopolitics and geopolitics. Despite the diversity and nuance in the puppets constructed in the narrative works analysed in Chapter 2, it was the duality of progressive objectification alongside sustained subjectification on both corporeal and psychological dimensions identified in Chapter 1 that generated the most powerful affective responses in the reader and it did so by capitalising on the contradictory connections between puppet and human conceptual categories. It is the puppet's capacity to evoke empathy even while the subjectivity and humanity of the character with whom we empathise are evacuated that speaks most clearly to the power of the connections forged across the topological gulf: The more numerous, varied and dense the connections, the stronger the contradictory combinatory form that results and the more potent the affective impact. Prompting a reconfiguration of certain aspects of the grotesque, it seems that bodies, borders, spaces, puppets and the grotesque are all mutually implicated in their co-construction.

These first two chapters, though, have focused overwhelmingly on fictional metaphorical uses of the puppet to establish an everyday, naïve understanding of the puppet. In Part 2, I progress to examine both more substantive uses of puppets and a broader range of cultural forms to dig more deeply into how puppets are understood and constructed and with what effects. Examining next the characterful use of puppets as either protagonists in their own right or actors within a performance, I extend this analysis to televisual and filmic uses of puppets and in Chapters 4 and 5, I explore musical and theatrical associations with and uses of puppets respectively. My disciplinary engagement also deepens as specific geographical concepts and interests are increasingly brought to bear on the puppet spatialities that emerge, with an increasingly tight focus on the conceptualisation and interrogation of puppet–human borders and practices of borderscaping.

## References

Anderson, B. (2012) Affect and biopower: Towards a politics of life. *Transactions of the Institute of British Geographers*, 37(1): 28–43.

Bakhtin, M. (1984) *Rabelais and his world*. Indiana University Press: Bloomington, IN.

Bicât, T. (2007) *Puppets and performing objects: A practical guide*. The Crowood Press: Marlborough.

Blomley, N. (2006) Uncritical critical geography? *Progress in Human Geography*, 30(1): 87–94.

Blomley, N. (2007) Critical geography: Anger and hope. *Progress in Human Geography*, 31(1): 53–65.

Blomley, N. (2008) Spaces of critical geography. *Progress in Human Geography*, 32(2): 285–293.

Connelly, F.S. (2012) *The grotesque in Western art and culture: The image at play*. Cambridge University Press: New York.

Duggan, R. (2016) *The grotesque in contemporary British fiction*. Manchester University Press: Manchester.

Foucault, M. (2007) *Security, territory, population: Lectures at the Collège De France, 1977–78*. Palgrave Macmillan: New York.

Francis, P. (2012) *Puppetry: A reader in theatre practice*. Palgrave Macmillan: Basingstoke.

Ghoddousi, P. & Page, S. (2020) Using ethnography and assemblage theory in political geography. *Geography Compass*, 14(10): e12533.

Gregory, D., Johnston, R., Pratt, G. et al. (2009) Geopolitics. In: D. Gregory, R. Johnston, G. Pratt, M.J. Watts & S. Whatmore (eds) *The dictionary of human geography*, 5th Edition. Wiley-Blackwell: Chichester, 300–302.

Gross, K. (2011) *Puppet: An essay on uncanny life*. University of Chicago Press: Chicago, IL.

Harpham, G. (1976) The grotesque: First principles. *The Journal of Aesthetics and Art Criticism*, 34(4): 461–468.

Harpham, G. (2006) *On the grotesque: Strategies of contradiction in art and literature*. The Davies Group Publishers: Aurora, CO.

Kayser, W. (1963) *The grotesque in art and literature*. Indiana University Press: Bloomington, IN.

Kitchin, R. (2014) Engaging publics: Writing as praxis. *Cultural Geographies*, 21(1): 153–157.

Lehtinen, A.A. (2009) Biopolitics. In: R. Kitchin & N. Thrift (eds) *International encyclopedia of human geography*. Elsevier (online resource), 320–323. Accessed 11 July 2021.

Murphy, A. (2006) Enhancing geography's role in public debate. *Annals of the Association of American Geographers*, 96(1): 1–13.

O'Tuathail, G. (1996) *Critical geopolitics: The politics of writing global space*. Routledge: London.

Philo, P. (2012) Foucault. In: P. Hubbard & R. Kitchin (eds) *Key thinkers on space and place*. SAGE: London, 162–170.

Reuber, P. (2009) Geopolitics. In: R. Kitchin & N. Thrift (eds) *International encyclopedia of human geography*. Elsevier (online resource), 441–451. Accessed 11 July 2021.

Rogers, A., Bear, C., Hunt, M., Mills, S. & Sandover, R. (2014). Intervention: The impact agenda and human geography in UK higher education. *ACME*, 13(1): 1–9.

Simonsen, K. (2013) In quest of a new humanism: Embodiment, experience and phenomenology as critical geography. *Progress in Human Geography*, 37(1): 10–26.

Simpson, P. (2017) Spacing the subject: Thinking subjectivity after non-representational theory. *Geography Compass*, 11: e12347.

Tillis, S. (1996) The actor occluded: Puppet theatre and acting theory. *Theatre Topics*, 6(2): 109–119.

Ward, K. (2006) Geography and public policy: Towards public geographies. *Progress in Human Geography*, 30(4): 495–503.

# Part 2
# Performative puppets

# Introduction to Part 2

Progressing from the naïve or everyday understanding of puppets established in Part 1, my attention in this part broadens in two directions. Firstly, it extends beyond a sole focus on literary representations to explore the uses of puppets in diverse cultural mediums. Secondly, it extends beyond metaphorical representations of the idea of the puppet to consider the use of puppets as material entities. Part 2 – *Performative Puppets* – addresses the substantial use of puppets as performers or characters in a range of cultural forms (literature, film, theatre and music) to move further beyond the identification of the significance of borders to conceptualise the generative potential of processes of borderscaping in crafting unique puppet spatialities. To that end, the question guiding this part asks what it is that grants puppets their performativity (their generative capability), but the analysis retains the perspective of the spectator rather than the performer, puppet maker or puppeteer and prioritises constructions of puppets over performance with puppets, although the two are clearly connected.

## Chapter summary

This part is constituted in three chapters. The first of these, Chapter 3, analyses works that make explicit use of puppets as either stand-ins for human actors, narrative devices or analytical themes. A detailed analysis of 50 works of popular culture (different to those used in Part 1) is undertaken, across film, television and literature, and covering a range of genres and both adult and child audiences. Through this analysis, a preliminary dimensional model of the uses of puppetry in popular culture is developed, which helps us to understand how puppets are used and to what effects. This analysis is then deepened to provide an account of the spatialities that are created in these works, covering the progressive puppetisation of everyday spaces, spaces of memory and spaces of the body to draw a distinction between fantastical spaces that are employed simply as a setting and the creation of peculiar – unique and strange – worlds through the interrogation of puppetness. These puppetised geographies are resituated in the disciplinary context by developing concepts from comic book geographies – borderscapes and the topological gulf – and integrating them with the grotesque to theorise an essential yet

DOI: 10.4324/9781003214861-8

mutually incompatible pairing of eliminating puppet/human distinctiveness and foregrounding their distinctiveness in a theatrical dissolution of the border. Chapter 3, then, both introduces a dimensional framework of puppet use and identifies the interrogation of the human/puppet border as a key site for the generation of unique puppet spatialities, which is conceptualised as a process of borderscaping.

In Chapter 4, my attention turns to the relationship between puppets and music to examine how puppets are represented or constructed in lyrics and music, from constructions of 'puppet' as a general category to specific puppet types (e.g., marionettes) or characters (e.g., Mr Punch). Given my personal lack of musical education, I situate my analytical method within disciplinary debates as to the need or otherwise for musical research to be undertaken by proficient or professional musicians, especially if it is based on the reception rather than the production of music. Two emphases emerge from this analysis, the first concerning the co-constitution of mood and movement and the second revolving around tensions and contrasts. As in Chapter 3, the musical analysis unearths further instantiations of the body grotesque, such as an evocation of anatomical disassembly, and the character contorted, as in the reconfiguration and redistribution of the anarchic character of Mr Punch, and further specifies the borderscaping that is done across varied topological gulfs in the musical construction of puppets.

Chapter 5 focuses on the use of puppets for specific dramatic and affective purposes, namely the conveyance of moral messaging and the generation of affective impact in theatrical productions. The chapter begins with a consideration of the potential within different styles of puppet that are visually similar but very different in operation. A comparison of the types of puppets used in The Muppets and *Avenue Q* reveals that the former precludes the creation of a body grotesque whereas the latter creates a very specific form of body grotesque through a combination of excessive faciality and a constitutive absence of legs, due to the different ways in which the puppets and puppeteers interact. Subsequently, a comparative investigation of the same story (*War Horse*; *Dr Dolittle*) in different contexts (novel, film, puppet play) deepens the critique emerging in this chapter by highlighting the limitations imposed on the creativity of a cultural work by using puppets merely as proxies for human and/or animal actors/characters. Two key areas for examination emerge from consideration of these works in the context of the quotidian understanding of puppets and the dimensional framework: The potential to use puppetness to bolster the moral messaging being fed through a story, and the power of employing puppetness in varied forms to generate dramatic effect and affective impact. This detailed comparative discussion of the divergent approaches to the borderscape draws out further developments in our understanding and conceptualisation of how puppetness is used for narrative and performative effect and provides a more specific formulation of the borderscape. Further, an explicit distinction is drawn between the border and the borderscape wherein the integrity of the puppet–puppeteer borderscape can be maintained even as the border at its heart is torn asunder, which is explored in relation to work on puppets as uncanny. This relationship between the border and the borderscape is proposed as a means of integrating slightly divergent views on the uncanny phenomenon variously called opalisation, double-vision or oscillation (Gross 2011; Francis 2012; Jurkowski 2013), which encapsulates a characteristic fluctuation between seeing a puppet as a living subject and seeing it as an inanimate object. Specifically, it is proposed that the borderscape acts as a scaffold that can maintain a sense of puppet unity even while the rupturing of the border destroys the illusion of vitality, making it easier to restore the illusion subsequently.

## Conclusion

Part 2 looks beyond performance with puppets to consider the significance of specific types of puppets for the worlds they can create and to examine the performativity of 'the puppet' to tease out how it can generate specific narrative and dramatic effects, create unique puppet worlds and evoke potent affective impacts. Overall, this part identifies a number of additional – performative – dimensions to puppet uses in cultural works, identifies different approaches to the treatment of the human/puppet border in generating bodies grotesque and characters contorted and establishes a distinction between the space of the border and its associated borderscape, which hold potential to interact for unique effects. The Conclusion to Part 2 pulls together the headline findings from these three chapters and outlines their significance for the discipline of geography, notably by further developing the dimensional framework to incorporate the outcomes of Chapters 4 and 5 and by picking up on affect as an undercurrent running through these chapters and relating this back to both the varied treatments of the border and the nature of the puppet as uncanny.

## References

Francis, P. (2012) *Puppetry: A reader in theatre practice.* Palgrave Macmillan: Basingstoke.

Gross, K. (2011) *Puppet: An essay on uncanny life.* University of Chicago Press: Chicago, IL.

Jurkowski, H. (2013) *Aspects of puppet theatre,* 2nd Edition. Palgrave Macmillan: Basingstoke.

# 3 Puppets in text and film

In this chapter[1], my analytical focus broadens both beyond metaphorical uses of puppets to consider their employment as characters or as proxies for human actors and beyond their literary use to consider the uses of puppets in filmic and televisual outputs, too. While film and literature have a long history of geographical engagement (Adams 2017), my comparative focus here is on cartoon and comic book geographies as these have recently been described as a newly identifiable area of cultural geography (Peterle 2017) and both comics/cartoons and puppets blur the distinction between high and low culture (Cutler Shershow 1994; Dittmer 2005), providing a degree of equivalence between the two cultural forms despite their differential treatment within academic circles, with comic but not puppet characters featuring in publications listed on the academic database Scopus (see Introduction for more detail).

Much social science work on comics and cartoons is geopolitically oriented, mostly at the national or international level (Dittmer 2005, 2007; Dunnett 2009; Thorogood 2016, 2018) but also within a metropolitan context (McCann 1999), with analysis revealing both hegemonic (Dittmer 2005, 2007) and counter-hegemonic (Dodds 2007; Thorogood 2016) narratives, and covering cultural outputs for children as well as adults (Buckingham 2007; Dunnett 2009). However, progressive diversification from this focus has been identified, with growing interest in the material form of the comic as explicitly geographical (Laurier 2014), generating variety in the spatial experiences in reading comics (Dittmer 2010; Dittmer and Latham 2014; Peterle 2017). At the same time, interest is growing in such cultural forms as methodologically informative in the context of both research (Dittmer 2010; Katz 2013; Thorogood 2018) and teaching (Kleeman 2006; Hammett and Mather 2011; Gomez 2014).

By contrast, very little geographical work addresses puppets explicitly and when puppets do feature, this is in a passing or metaphorical fashion, as in labelling post-conflict states 'puppets' if they remain tied to a patron state (Bakke et al. 2018). Specific puppet figures have occasionally featured in such geopolitical work, either to establish a genre lineage from the Punch

DOI: 10.4324/9781003214861-9

magazine (Kleeman 2006) or comparing the comedic vulgarity of *South Park* with *Spitting Image* (Thorogood 2016), but puppets as a cultural form or geographical phenomenon are not directly interrogated. Finally, although there has been some experimentation with puppets in a teaching context (Pelletti 1973; Tanner 2016), these examples are also rare. Consequently, if comics/cartoons are described as being under-researched in geography (Robson 2019), puppets are utterly unappreciated.

As outlined in the Introduction, puppets are not confined to children's entertainment but are also used in popular adult cultural outputs, nor are they confined to any one cultural medium or genre and attending to this variety can generate new perspectives on puppet use and effects. Building on the analysis of metaphorical uses of puppets in Part 1, I focus here on the employment of puppets either as actors or characters – as material entities – not just as an idea. This chapter focuses on representations of puppets in films, television programmes and literature covering genres from humour to horror and catering for both child and adult audiences. Table 3.1 presents the 50 works analysed, which were identified through existing knowledge, internet searches and suggestions received during conversations with others. While numerous other puppet works could have been included (e.g., *Spitting Image*, *Muffin the Mule* and so on), the sample is sufficient to generate a sense of how puppets are treated in each category, and some puppet works have been reserved for analysis in specific contexts (e.g., *Spitting Image* for political analysis). Each work was analysed individually to identify the main representations, associations and uses of puppets, before two stages of comparative analysis were undertaken. The within-group comparative analysis considered similarities and differences between texts that had been grouped according to audience and type (e.g., adult films, child fiction), which, alongside analysis of similarities and differences between groups, generated an analytical landscape of the uses of puppetry in popular culture.

This chapter explores how puppets are used in these diverse cultural outputs and to what effect, generating a dimensional model of these uses to stimulate further methodological and analytical work. Subsequently, the analysis is extended by examining puppet spatialities and temporalities (how puppets are generative of their own spatial and temporal realities), which in turn is employed to advance geographical conceptualisations in existing work on comics and cartoons. The discussion culminates in the specification of intercorporation as the way in which puppet and human bodily forms are brought together, which is both related to and distinguished from performative ideas of transembodiment (the transfer of capacities between puppeteer and puppet), the notion of the topological gulf in comic book geographies (across which readers must forge their own narrative continuities) and the grotesque. Finally, the conclusion draws out the key points emerging from these analyses and highlights their disciplinary significance.

*Table 3.1* List of works analysed

| | Adult | Child |
|---|---|---|
| Film & TV | Dolls (S. Gordon, 1986/2014)<br>Robert (A. Jones, 2015)<br>Dead Silence (J. Wan, 2007)<br>Puppet Master (D. Schmoeller, 2002)<br>Being John Malkovich (S. Jonze, 2003)<br>Team America (T. Parker, 2004)<br>The Happytime Murders (B. Henson, 2018)<br>Gone Girl (D. Fincher, 2014)<br>The Game (D. Fincher, 2001)<br>The Beaver (J. Foster, 2011)<br>Murder She Wrote – Where Have You Gone, Billy Boy? (J.L. Moxey, 1991)<br>Murder She Wrote – Something Foul in Flappieville (A.P. Shaw, 1996)<br>Friday Night Dinner – Lord Luck (R. Popper, 2018)<br>Endeavour – Neverland (G. Sax, 2014)<br>Endeavour – Apollo (S. Evans, 2019)<br>Midsomer Murders – Destroying Angel (D. Tucker, 2001)<br>Dr Who – The Time of the Doctor (J. Payne, 2014)<br>Dr Who – The Snowmen (J. Payne, 2014)<br>Murdoch Mysteries – Belly Speaker (F. Mann, 2008)<br>Hold the Sunset – season two (S. Johnson, 2019) | Goosebumps (R. Letterman, 2015)<br>Thunderbirds Are Go (D. Lane, 1966)<br>Pinocchio (N. Ferguson, 1940)<br>The Muppet Christmas Carol (B. Henson, 1992)<br>The Great Muppet Caper (J. Henson, 1981)<br>Muppet Treasure Island (J. Henson, 1996)<br>Muppets from Space (T. Hill, 1999)<br>The Muppet Movie (J. Frawley, 1979)<br>Muppets Most Wanted (J. Bobin, 2014)<br>Labyrinth (J. Henson, 1986)<br>Dark Crystal (J. Henson, 1982)<br>The Muppet Show (J. Henson, 2002)<br>Furchester Hotel (dir: various, 2014-2017) |

Fiction

*The Death of Mr Punch* (J. Carter, 2016)
*The Comical Tragedy or the Tragical Comedy of Mr Punch* (N. Gaiman, 2015)
*Rivers of London* (B. Aaronovitch, 2011)
*Riddley Walker* (R. Hoban, 2012)
*Pinocchio* (C. Collodi, 1862)
*Bryant and May and the Memory of Blood* (C. Fowler, 2011)
*The Old Curiosity Shop* (C. Dickens, no date)

*Night of Cakes and Puppets* (L. Taylor, 2017)
*The Magicians of Caprona* (D. Wynn Jones, 2008)
*Dodger* (T. Pratchett, 2013)
*The Box of Delights* (J. Masefield, 1935/2014)
*Pinocchio by Pinocchio* (M. Morpurgo, 2013)
*The Invisible Detective: The Paranormal Puppet Show* (J. Richards, 2003)
*Goosebumps: Night of the Living Dummy* (R.L. Stine, 2015)
*Goosebumps: Revenge of the Living Dummy* (R.L. Stine, 2008)
*Goosebumps: Night of the Living Dummy 3* (R.L. Stine, 1996)
*Goosebumps: Night of the Puppet People* (R.L. Stine, 2015)

## Puppets in popular culture

Across the works analysed, puppets are utilised to different degrees and in diverse ways, ranging from minimal use as a simple object, through moderate use as a proxy for humans and metaphors for human life, to a core topical focus through which the human condition is interrogated for narrative purposes. These can be organised into a series of dimensions (utilitarian, relational and qualitative), as outlined in Figure 3.1.

The utilitarian dimension relates to the functional role of puppets and the extent of their use. Some cultural outputs use puppets simply as a material object or as a proxy for a human actor, while others present them as autonomous agents, and yet others employ them as a narrative tool to question what it is to be human rather than puppet (e.g., exploring issues of control in *Being John Malkovich*). Similarly, puppets can be used either minimally or extensively and either focally or incidentally. Both *Riddley Walker* and *Dodger* draw on Punch and Judy, but the former makes much fuller use of the show as a mechanism for maintaining governmental control of a dystopian society than the latter's analogical use of the same show to explore interpersonal violence. Even this first dimension brings myriad combinations of the function and extent of puppet use for academic exploration.

The second dimension is relational, and addresses both individual and social relations between humans and puppets. At the individual level, this relationship is sometimes framed as one of interaction, as in children's fiction posing questions as to who or what is controlling the puppet that seemingly acts on its own. In other examples, it is more a case of interchangeability of

| Utilitarian | Role | Incidental | Indirect | Focal |
|---|---|---|---|---|
| | Proxy | Puppets rarely feature | Puppets and humans used similarly | Puppets used instead of humans |
| | Narrative | Puppets as unquestioned objects | Puppets provide new perspective but not intentional or central | Puppets questioned as a narrative device |
| | | *Increasing integration* ⟶ | | |
| | Extent | Minimal | Partial | Full |
| | | Puppets not a common or major aspect of the work | Puppets feature significantly but not a/the driving force | Puppets used extensively and thoroughly, the main feature |
| Relational | Individual | Informational | Social | Psychological |
| | | Puppet is simply part of the world, factually. | Puppet as social mediator, modifying social acceptability | Puppet as communicative facilitator |
| | | *Increasing interrogation* ⟶ | | |
| | Social | Discrete | Metaphorical | Constitutive |
| | | Puppet and human societies distinct | Puppet society/show used as metaphor for human society/life | Puppets humanised, humans puppetised |
| Qualitative | Puppetness | Negligible | Moderate | Substantive |
| | | No consideration of puppet qualities/nature | Some consideration of puppet qualities/nature | Significant consideration of puppet qualities/nature |
| | | *Increasing intercorporation* ⟶ | | |
| | Ontology | Conventional | Creative | Transformational |
| | | Narrative remains within human context | Some modification of or addition to human context | New worlds created by merging puppets and humans |

*Figure 3.1* Dimensions of puppet use in popular culture.

Source: Banfield 2020.

human and puppet, although this interchangeability is managed differently in different works. In *Rivers of London*, magic is employed to enable the ghost of a Punch and Judy Professor to exact revenge by turning the physical form of his victims into Mr Punch, whereas *The Death of Mr Punch* adopts a psychologised perspective as a former Punch and Judy Professor acts as if he is Mr Punch as he succumbs to dementia. Sometimes these relations are set within a social context, perhaps being used as an object but for a social purpose, as in *Murder She Wrote – Something Foul in Flappieville*, which utilises a puppet for educational purposes to explain the world to young children. Alternatively, the puppet is framed as socially situated, whereby the puppet is viewed differently by different people. In *Friday Night Dinner – Lord Luck* the puppet makes socially excusable utterances that would be unacceptable if said by a human, thereby mediating standards of social acceptability. Then again, *Murder She Wrote – Where Have You Gone, Billy Boy?* presents the puppet as enabling the puppeteer to say that which would otherwise remain inarticulable not because of social mores but psychological distress. Towards this end of the dimension, puppets become increasingly significant social and psychological actants. At the individual relational level, the location of communicative autonomy shifts from the puppeteer to the puppet as the communicative role played by the puppet changes from informational dissemination through social mediation to psychological facilitation.

At the social level of the relational dimension, puppets are used to comment on human society through comparison or integration of human and puppet communities. Three examples employing Punch and Judy in children's fiction exemplify this. In *The Box of Delights*, one of the main characters is a Punch and Judy Professor but the show itself only receives fleeting mention. Instead, a dual social context of human and magical communities is established, each with its own social norms, which acts as a mirror to the other. In *Dodger* the social appropriateness of Punch and Judy is explored, drawing parallels between the puppet show and social life in the use of violence for the purposes of control. By contrast, *The Magicians of Caprona* involves the puppetisation of people by magical means to inflict violence on them by performing the Punch and Judy show, rather than the show metaphorically standing for the realities of social life. Here, puppet and human communities progress from being discrete to being metaphorically related to being mutually constitutive in the puppetisation of humans and the humanisation of puppets, again reflecting varying degrees of human–puppet integration but at a societal level.

The third dimension relates to the qualities of puppets and includes consideration of puppetness and ontology (world-making potential). The extent to which 'puppetness' as a quality is an identifiable feature or focus of the work is often related to puppet role. In *Thunderbirds Are Go*, puppets are used as proxies for human actors but their nature as puppets and the opportunities this brings for plot and action are not directly explored, despite footage of human hands being used for close-up scenes interchangeably with puppet performance for whole-body scenes. By contrast, *Pinocchio* explicitly questions the human condition through the material form, more-than-human sociality and subjectivity of Pinocchio. Between these two are The Muppets, especially *The Happytime Murders*, the first Muppet movie for adults. Here, human and puppet communities live together but discrimination is rife due to their varied materialities: Fleshy humans and fluffy puppets. The puppets are constructed as suspended between objectivity, with insults denigrating them as fluffy socks, and animality, with laws prohibiting the poaching of puppets for their feet, so some use is made of puppetness. Moreover, the human pro-tagonist has received a puppet liver transplant, which is used to suggest that she has gone soft on puppets due to her new liver being fluffy rather than fleshy, blurring the distinction between humanness and puppetness at the level of the body. The puppets, though, acknowledge that they live in a human world, so the world-making (ontological) possibilities are limited to a human context as humans are precluded from participating in a puppet world (what-ever that might look like), and the two communities are held apart by their mutual prejudices and separate moral codes. However, although puppetness is related to role, there is not a direct correspondence between basic/inciden-tal role and negligible puppetness on the one hand and focal role and sub-stantial puppetness on the other. *Team America* makes only basic use of puppets as human proxies but makes moderate use of puppetness for come-dic (highlighting what puppets cannot do) or sensationalist (killing puppets in as many gory ways as possible) purposes. By contrast, *The Happytime Murders* are primarily populated by puppets and puppets are used as a sub-stantial narrative mechanism so are towards the focal end of the role dimen-sion but the interrogative use of puppetness, although in evidence, is only moderately developed.

Finally, and related to puppetness, is the extent to which the use of puppets affords alternative, unusual and incredible world-making possibilities. While the extendable arms of Gonzo (*The Muppet Show*) or a puppet's ability to vomit leeches at will (*Puppet Master*) is suggestive of more than human capa-bilities, these are still often set within a human context, whether a theatre (*The Muppet Movie*), international terrorism (*Team America*) or a domestic environment (*Dolls, Robert*). In some instances, this human context is itself puppetised, as in the modification of classic literature to fit the puppet char-acters (e.g., two Marley brothers in *The Muppet Christmas Carol* to cater for Statler and Waldorf – the grumpy old men from the balcony), but in its most extreme form, whole new possibilities of existence come into being (e.g., John Malkovich's ability to enter his own head).

Importantly, it is the ability to sustain and integrate an element of human-ness and puppetness that brings greatest potential, as each is liberated from the constraints of their conventional material form and existential possibili-ties. However, this is complicated by the extent to which puppets are used and puppetness is employed. If puppets are used too simplistically, they are con-ceived either as objects or as proxies for humans, and if too much emphasis is placed on the human, the more-than-human puppet capacities are neglected, meaning that the existential possibilities are confined to human contexts (unless another factor is introduced, such as magic). For example, *Rivers of London* generates fantastical temporalities and subjectivities through magical means as the ghost of a Punch and Judy Professor haunts the present and the contemporary characters travel back through history in pursuit of the ghostly felon, while *The Death of Mr Punch* generates disorienting temporal and existential conditions through the human experience of dementia. However, *Dolls* and *The Magicians of Caprona* both conduct a dualistic transposition of human-being-puppet and puppet-being-human, in a manner that goes beyond that achieved by *The Happytime Murders*. What it is to be either human or puppet is unclear; the potentialities for action, experience and autonomy of each are transformed; and moral and ethical as well as existential questions arise. Developed to its fullest extent, this dimension generates a whole new world of human–puppet materiality, sociality, practicality and morality. Humanising puppets can be very powerful and – in *The Death of Mr Punch* – very poignant, but simultaneously puppetising humans is where the greatest potential lies for the creation of unique worlds that interrogate and optimise puppetness for narrative purposes.

These dimensions characterise how puppets are constructed and used in popular culture and to what effects. Two dimensions are concerned with utility: Their role and the extent to which they are used. A further two are concerned with relationality, at both the individual and collective level. The final two dimensions (puppetness and ontology) attend to qualitative aspects of puppets; questioning the nature of the puppet and creating new worlds. It is entirely possible that further dimensions and distinctions can be drawn through subsequent research, as different aspects of puppetness might vary in significance between puppet types (marionettes, glove puppets), and – indeed – later in Part 2, this framework is developed further. The relations among the dimensions are also worth exploring further as it is the role rather than the extent of the use of puppets that is most important in determining the effects generated: A focal role for questioning what a puppet is, prompts engagement with puppetness, but this is not inevitable as the question of what a puppet is, might be answered by using the puppet solely as an object, precluding any such potential. Consequently, this analytical framework facilitates both understanding and investigating how diversely puppets are used in cultural representations and in the creation of puppet spaces and worlds, although it is developed further in the subsequent chapters in Part 2.

## Puppet geographies

In this section, I explore puppet geographies through three lines of discussion: Puppet bodily spaces, which take diverse forms and merge with each other to different degrees; spaces of puppet memory, which are found to be more extensive and enduring than human memory; and puppet spaces and worlds, which considers the constellation of factors with greatest potential to generate unique worlds that are peculiar to puppets.

### Puppet bodily spaces

The ways in which puppet and human (puppeteer) bodily spaces are employed vary considerably across the works analysed. Sometimes the puppet has a mind of its own within a yoked dual persona, fluctuating between being an external other and an internalised aspect of their human operator, and *Pinocchio*, of course, is the archetypal example of puppet–human interchangeability in wanting to become, and becoming, a real boy. In horror, a curse is sometimes transferred from puppets to humans, while a human spirit can be incorporated into a puppet in a metaphysical merging rather than a transition from one to the other. Intercorporeal merging of puppet and person also features. In *Dolls*, the elderly residents of a remote house administer justice to visitors lacking sufficient reverence for toys by transforming them into dolls, and the man who acts violently towards a Punch puppet is turned into a figure of Punch, complete with a hooked nose and jingly hat. Here, the bodily forms become one, but the subjectivities remain distinct: Punch remains a material object despite being enlivened with human vitality, but the human is aware of their own captivity within the material form of Punch. *Rivers of London* takes a simultaneously corporeal and spiritual approach. The ghost of Henry Pyke (aka Punch) uses real people to enact his own show by possessing their body, which visibly transfigures as the chin bulges and the nose stretches. Pyke can both sequester other people's bodies and transition between the physical forms of Pyke and Punch, rendering indistinguishable the subjectivities of puppet and human and unifying their bodily forms and appearances.

Intercorporeal blending of a different sort is evident in *The Death of Mr Punch*, in which a former Punch and Judy Professor increasingly confuses his own identity as the Professor and the identity of the Punch with which he used to work as he succumbs to dementia. Dressing as Mr Punch, waving his slapstick and adopting Punch-like behaviours, the Professor increasingly embodies the puppet with which he used to perform. *Being John Malkovich*, though, provides the most extreme example here in the physical passage of one person into the brain of another. When his head is occupied by someone who uses their control of John Malkovich to engage in sexual intercourse with a third person, the love-making entity involves three personae but only two bodies. The occupier's body is inside John Malkovich's brain, but John Malkovich's subjectivity is overridden rather than replaced as he expresses awareness of being used as a mouthpiece by someone else. Bodies, then, can be merged just as much as subjectivities and in equally diverse ways, and the space of the conjoined human–puppet entity forms a more explicit focus in works that engage more thoroughly with puppetness. Unlike treating puppets merely as objects or human proxies, which precludes any engagement with intercorporation (the integration of corporeality between human and puppet), the human–puppet bodily border becomes the target for narrative creativity. This allows the transference of human qualities and capacities to the puppet and vice versa, unsettling assumptions as to what constitutes 'puppet' and 'human' by liberating them from the constraints of their respective material forms to create peculiar intercorporeal capacities.

## Spaces of puppet memory

One perhaps surprising such capacity is memory. A common trope of puppets, especially ventriloquists' dummies, is that they retain knowledge and could tell of things that humans would not wish others to know, and sometimes that they do not know themselves. It is often in murder mysteries that puppet memory and the risk of or need for its disclosure features, as the puppet holds vital information to solve the crime, sometimes over extensive time periods. In *Midsomer Murders – Destroying Angel*, however, Mr Punch has a tradition of revealing local scandals, positioning the puppet as a barometer of both truth and (ironically) social probity. This is developed further in *Bryant and May and the Memory of Blood*, in which the Punch story is re-enacted through modern-day murders in revenge for familial wrongdoings. Moreover, the ability of puppets to facilitate human articulation of otherwise inarticulable events and experiences – even against the wishes of their human operator – not only evidences the sharing of memory between human and puppet but also the puppet's power to resist any psychological efforts on the part of the human to suppress that memory. In this example of the psychological individual relational dimension, puppets have a clarity and continuity of memory that surpasses human memory, allied with greater communicative autonomy, enabling truths to be revealed and preventing misdemeanours – both legal and moral – from being unjustly neglected.

Social memory, too, can be inscribed in and reconstituted through puppets, as exemplified by the governmental use of the Punch and Judy show in *Riddley Walker*. Similarly, in *Bryant and May and the Memory of Blood*, Mr Punch is considered to demonstrate, perform or explain life, illustrating the unpalatable promise and profitability of opportunism in the making of the modern world, while in *The Comical Tragedy or the Tragical Comedy of Mr Punch*, puppets conscript children into social practices and norms as carriers and communicators of social memory. In *Rivers of London*, Punch describes his spirit as irrepressible, suggesting the inevitable perpetuation of Punch's take on society; a perpetuity supported and continued by the frequency of his appearance in the works analysed here, at least 15 of which feature the irrepressible Mr Punch.

Puppets are also used to explore the fallibility of human memory, most clearly demonstrated in *The Death of Mr Punch* and *The Comical Tragedy or the Tragical Comedy of Mr Punch*. The first narrates the crumbling and confusing of memory and identity for a former Punch and Judy Professor in the grips of dementia, who lives as a human in the present as if he is the puppet from his past, with jumbled associations and juxtaposed recollections. The second revisits childhood memories of the protagonist's Grandfather's performances, with similarly jumbled and juxtaposed memories, associations and emotions. Both texts highlight the duplicity of memory as the characters shift between past and present, reality and fantasy, dream and memory. In the former, a model of Mr Punch is used to try to help the protagonist reconnect with his past as a material embodiment of the memories that are increasingly muddled, while in the latter the material space of a Punch and Judy proscenium conjures memories of other times and places that involve similar materialities. These two texts work well together in addressing issues of memory, loss and death in a similar way – through the materialities of Punch and Judy – but from different perspectives: The performer in the former and the spectator in the latter. Puppets, then, are diversely and intimately bound with memory, whether as a reservoir of elusive truths, a medium for the perpetuation of social memory, or a material repository of memories. They can both conceal and reveal truths and secrets, and can both remember more, for longer than humans and empower (or overpower) their human operator in communicating memories. Puppets remind us that memory, like subjectivity, is not confined within the body but is distributed in more-than-human entanglements and that memories, like puppets, shift between reality and fantasy, the normative and the pathological.

### Puppet spaces and worlds

The spaces and places that feature in the analysed works are many and varied, and a spatial continuum can be discerned ranging from a focus on familiar spaces (everyday and iconic sites) and repeated engagement with specific spaces (attics, bedrooms), through the reworking of these sites for puppet purposes (perhaps in terms of scale or functionality) to the generation of fantastical places that are even less recognisable within human frames of reference. Among everyday spaces, specific types of space feature more commonly than others. Dark, high and low spaces, such as attics, cellars and abandoned buildings, suggest a strong association with horror. However, puppets in these works are often found rather than purchased so cellars and attics might be as much a trope of puppets as of horror. Moreover, specific domestic spaces become important. While kitchens as a source of lethal weapons again suggests a horror connection, bedrooms evoke privacy and trust, with the puppet granted privileged access to an inner sanctum where the human–puppet relationship is secured prior to the manipulation of the human. Certain everyday spaces, then, become preferentially associated with puppets, while they can also be puppetised in generating spaces specific to that puppet form and story.

There are also distinctions to be drawn between works that generate their own fictional world and those that remain within human parameters, and between fantastical places created as settings and internal fictional worlds created through the integration of puppet and human. For example, *Labyrinth* and *Dark Crystal* involve fantastical places but these are not generated through interrogation of puppetness: The fantastical places are settings and the puppets are simply characters in the story. By contrast, the internal fictional world of *Being John Malkovich*, and to a lesser degree *Dolls* and *The*

*Magicians of Caprona*, springs from the attention directed towards puppetness. *Being John Malkovich*, for example, involves a hidden passage that serves as a portal between an office environment and John Malkovich's brain, enabling characters to enter his brain and control him like a puppet. Here, the office would be the fantastical place as the ceiling is so low that most people are forced to navigate the space in a stooped posture, but the internal fictional world is the tunnel-brain nexus that enables the puppetised control of John Malkovich. Consequently, the use of spaces and places in these works is distinct from the new worlds that are sometimes created through an interrogation of puppetness, even while they might contribute to the creation of that new world.

Spaces in these works are multiple and diverse, and Figure 3.2 visualises these complex puppet spaces from iconic sites to fantastical places, highlighting both everyday spaces associated with puppets and the significance of bodily spaces and spaces of memory across the puppet–human border. Some spaces (attics) encapsulate the ambiguity of the puppet's past, while others are important in nurturing the puppet–human relationship of dependency (bedrooms). Bodily spaces are important in enabling the transference of qualities and capacities across the puppet–puppeteer bodily border, while spaces of memory are important in both rendering fluid past and present, reality and fantasy and unsettling assumptions of relative autonomy and probity. It is through explicit interrogation of puppetness that the most peculiar puppet worlds are generated, and as space, the body and memory are all bound up with puppetness, the prime space for the creation of such worlds

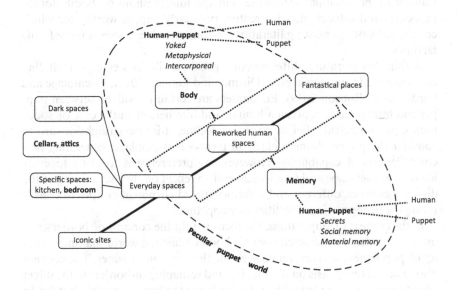

*Figure 3.2* Puppet spatialities.

Source: Banfield 2020.

lies in the intersection between variously conflated spaces, conjoined bodies and contrary memories across the border between puppet and puppeteer.

## Intercorporation

In this final section, I develop more thoroughly the notion of intercorporation by engaging with work in performance studies on 'transembodiment' between puppets and puppeteers, and – after establishing how intercorporation differs from this – returning to geographical work on comics and cartoons to ground intercorporation in this body of work.

Transembodiment, a term used in performance studies to denote the transfer of embodied techniques among puppeteers and puppets, involves both direct techniques that are visible and knowable (e.g., gesture, rhythm) and indirect techniques that are not visible (e.g., drawing on memory and emotion), which together imbue the puppet with a sense of weight, liveliness and character (Mello 2016). The bidirectional transfer of these techniques between animate and inanimate performers suggests strong synergies with the idea of intercorporation proposed here, but my interest in this chapter is not with the practical doing of puppetry but with the uses to which puppets are put and the interrogation of that puppet–puppeteer relationship for narrative creativity rather than its utilisation for a convincing performance. Consequently, while I also attend to the puppet–puppeteer relationship as a specific example of the puppet–human relationship, I – for this volume – maintain the perspective of the spectator and probe the employment of puppets rather than performance with puppets. Intercorporation is not about the transfer of professional techniques but the intermeshing of bodily forms, capacities and subjectivities in creating peculiarly puppet worlds, for which conceptual work in existing literature on comics and cartoons is more fertile territory.

Within this literature, different conceptual possibilities emerge, including encounter and topological gulf (Dittmer and Latham 2014), assemblage and borderscape (Robson 2019). Encounters are certainly evident between puppet and human, and between a human and him/herself in the case of social mores, psychological relief or cognitive decline. It is also straightforward to consider the puppet–human entity in terms of assemblage, generating new constellations of capabilities. However, my preference is to bring together ideas of borderscape and topological gulf to reflect on the overcoming of the distance between different and de/reformulated corporeal entities in constituting the ontogenetic capabilities of puppetness.

In the context of puppetness, the focus within the concept of borderscape on state borders is redirected towards bodies while the world-making capacity of puppetness remains reminiscent of the creation of states. The concept also retains its emphasis on the shaping and reshaping of borders in the direct use of puppetness to play with and reconfigure the human–puppet border. In terms of the dimensional framework (see Figure 3.1), such borderscaping might feature within the focal narrative element of the utilitarian dimension and within the psychological individual relational dimension and would characterise the 'upper' end of the qualitative dimension, finding maximum expression in substantive puppetness and transformational ontological categories. This pinpoints which parts of which dimensions allow peculiarly puppet worlds to emerge, but this is contingent upon the ways in which diverse elements – puppet, human, plot, genre and so on – come together. Consistent with the borderscape as constituted by different elements working together (Robson 2019) in representing the area around a border (dell'Agnese and Amilhat Szary 2015), the area represented in the particularities of the work created needs to be experienced as real in the performative consumption of the work. However, rather than distinguishing between radical dissolution and theatrical strengthening of the border as alternative approaches to reshaping it (van Houtum and Eker 2015), the interrogation of puppetness generates the *theatrical dissolution* of the human–puppet border. This bodily border between puppet and puppeteer is dissolved by unsettling distinctions that are usually drawn between them, but it is dissolved in such a way that attention is sustained on the very distinction that has been dissolved. It is in this sense that the dissolution is theatrical, as the peculiar capacities, bodily forms and spaces that are created through it throw light back onto those very distinctions. The peculiarity of the world created sustains our attention on the very border that has been dissolved in creating that world. The border, then, is simultaneously dissolved and reinscribed.

This affords an inverted appreciation of borders explored in work on comic book geographies, in which the gutter is considered a topological gulf or anti-optical void requiring the reader to forge connections between successive panels in the story (Dittmer 2010; Dittmer and Latham 2014), as the human–puppet relations are forged by the creator rather than the reader/viewer. Here, the challenge for the reader/viewer is not to forge their own connections but to invest in the intercorporation that has been constructed against a background of retained recognition of the distinction between human and puppet. The intercorporation is given in that it is determined by the creator, but it is also uncertain and unstable, requiring constant active maintenance in the performative consumption of the work. Cognitively, the void between human and puppet is maintained by the impossibility of the intercorporation that is presented, yet the intercorporation denies that void on a performative level. In works engaging directly with puppetness, the spectator is challenged not with overcoming an anti-optical void but with investing in the intercorporation presented while the distinction between puppet and human is kept within focus in the very interrogation of human–puppet relations that generate the intercorporation in the first place. This torturously oxymoronic process simultaneously sustains and eliminates human–puppet distinctiveness across the human–puppet bodily border by overcoming that distinctiveness only through interrogating it. Ultimately, then, we can think about the spaces and worlds generated by an interrogation of puppetness both in their individual peculiarity and in this peculiar practice of borderscaping.

The distinction drawn out here between the forging of connections between different elements across a topological gulf in comic book geographies and the investment in the intercorporation through the theatrical dissolution of the human–puppet border brings us to the significance of work on the grotesque, especially the body grotesque. With intercorporation, there is no anti-optical void across which the reader must forge connections, as the distinctions and the border remain very much within view. There does remain a gulf, however, because the impossibility of dissolving the border between the distinct elements that constitute the border also remains evident. Crucially, the dissolution of the border removes any notion of difference in terms of distance and places the emphasis firmly on the coterminous existence of the distinctions in one puppet–human bodily form, establishing this gulf as topological rather than topographical. The relations between the distinct elements are more important than the differences between them even though those relations are relations of difference (contradiction).

The erasure of borders between – or the bringing together of – incongruous elements is exemplary of the grotesque (Kayser 1963; Harpham 1976, 2006; Danow 1995; Goodwin 2009; Connelly 2012) and explicit engagement with the border between bodies, often resulting in the deformation or exaggeration of bodily features and capacities, is central to the body grotesque (Harpham 1976; Bakhtin 1984; Duggan 2016). Thus, the coterminous existence of different and incongruous bodily forms in one entity (the puppet–human form) is exemplary of the body grotesque. From the examples discussed here, at least three distinct forms of body grotesque can be identified. The first is inhabitation, whether through the possession (*Robert*), imprisonment (*Dolls*) or occupation (*Being John Malkovich*) of one bodily form by or in another entity, although this does not always result in an obvious body grotesque. With *Robert* and *Being John Malkovich*, for example, the bodily appearance of the puppetised form does not change, but the puppetised characters in *Dolls* are visibly a composite form between person and puppet. There is, then, a distinction to be drawn between the body grotesque as form and the body grotesque as appearance. The second is transfiguration, which does generate an obvious body grotesque and is most evident in the stretching of the nose and chin of the sequestered victims in *Rivers of London*, and which is an example of classic grotesquery in its targeting and exaggeration of specific facial features (Bakhtin 1984). The third is integration, as in the surgical incorporation of a puppet liver into a human character in *The Happytime Murders*, which might but need not generate a visibly obvious body grotesque.

At the same time, and picking up the thread from Chapter 2, in which the twin notion of the character contorted was proposed, there are also at least three forms of character contorted that emerge from this chapter. The first is modification, typically seen in the modification of a character's behaviour as they exhibit their puppetised form, such as in the direction of action through the force of magic (e.g., *Magicians of Caprona*). The second is conflation and is best exemplified in the bringing together – whether collaboratively or combatively – of puppet and human memories or communicative abilities (e.g., *Endeavour – Neverland*; *Murder She Wrote – Where Have You Gone, Billy Boy?*). Finally, the third is identification, as in the muddling of human and puppet identities in the grips of dementia (*The Death of Mr Punch*). While this typology of bodies grotesque and characters contorted might not encompass all the ways in which puppets are constructed in popular culture discussed in this chapter (e.g., as a repository of social memory), it does provide specificity as to diverse ways in which the body grotesque (and/or character contorted) is constituted through constructions of puppets, each of which stimulates its own unique narrative potential through its generation of particular and peculiar puppet–human intercorporealities, and each of which is firmly grounded in the explicit interrogation of puppetness as a practice of borderscaping.

## Conclusion

The analysis in this chapter teased out six dimensions – utilitarian (role, extent), relational (individual, social), qualitative (puppetness, ontology) – found to characterise different approaches to the use of puppets. Collectively, these provide an analytical framework for subsequent research that might productively look in more detail at either the intersections between cultural forms, practices, media and genres or the different forms and arrangements of puppetry to refine these dimensions further.

A detailed investigation of space in these works identified a preponderance of certain everyday spaces, alongside a graduated reworking of everyday spaces into spaces of puppetness as the nature of a puppet and the human–puppet relationship are increasingly employed as focal factors in the telling of the story. Moreover, the significance of bodily spaces and spaces of memory was highlighted in the generation of these peculiarly puppet worlds, whereby the most interesting puppet worlds emerge from the integration of puppet–human conjoined bodies, conflated spaces and conflicted memories. The notion of intercorporation was proposed to denote this integration, which was compared to and distinguished from performative ideas of transembodiment. These puppetised geographies were resituated in the disciplinary context by developing and integrating concepts from comic book geographies: Borderscapes were redirected to the bodily border between puppet and human, and the forging of relations across the topological gulf was compared to the investment in the intercorporation. These were brought together in the specification of intercorporation as a theatrical dissolution of the puppet–puppeteer/human border through the simultaneous elimination and reinstatement of puppet–puppeteer/human distinctiveness.

This, in turn, was brought into conversation with the grotesque and a typology of three forms of body grotesque and character contorted evident in the examples discussed in this chapter was proposed, which provides an alternative yet related conceptualisation of the diverse constructions of puppets. While intercorporation is distinguished from the forging of connections across an anti-optical void in comic book geographies, it remains associated with such ideas through the recognition of a topological gulf, which is a gulf in terms of the impossibility of bridging it but topological in that it is bridged through the theatrical dissolution of the categorical and bodily boundaries that constitute it. Similarly, while there are clear synergies between puppets and the (body) grotesque and three distinct forms of body grotesque have been specified, the notion of intercorporation (which also encompasses the character contorted) is also distinct from the grotesque as it allows for less extreme or less excessive formulations than would be commonly associated with the carnivalesque character of the grotesque. Excess, hyperbole, exaggeration and degradation are all characteristic qualities of the grotesque (Bakhtin 1984; Li 2009; Duggan 2016) and all can characterise puppet constructions, but they need not do so. In other words, while puppets are classic examples of the grotesque and can constitute their own body grotesque, not all puppets are bodies grotesque in appearance, even though they are in form. Intercorporation – as the mechanism identified here through which the most creative and peculiar puppet worlds are brought into being – accommodates this distinction, as intercorporation can – but need not – generate a visible body grotesque.

This chapter, then, both provides a dimensional framework that characterises the varied uses of puppets and the effects that they generate, with the potential to stimulate and guide further scholarship, and conceptualises the means by which the most peculiar puppet worlds are generated in the notion of intercorporation as a theatrical dissolution of the human–puppet bodily border, with potential to inform our conceptual understanding and analytical use of more familiar ideas of the borderscape, the topological gulf and the (body) grotesque. Puppetness is a spatial concern anchored on a bodily border, and the spatialities generated by these explorations of puppetness, especially in their more extreme forms, are certainly peculiar, in both senses of being unique and being strange.

## Note

1  This chapter was first published in *Cultural Geographies* (see Banfield 2020), although some of that paper's introductory material has been relocated to the Introduction to this volume. In addition, the conceptual engagement has been expanded to incorporate the grotesque, and the structure, narrative and referencing have been modified slightly to ensure consistency throughout this work. Otherwise, the original work is substantively unchanged.

## References

Adams, P.C. (2017) Geographies of media and communication 1: Metaphysics of encounter. *Progress in Human Geography*, 41: 365–374.

Bakhtin, M. (1984) *Rabelais and his world*. Indiana University Press: Bloomington, IN.

Bakke, K.M., Linke, A.M., O'Loughlin, J. & Toal, G. (2018) Dynamics of state-building after war: External-internal relations in Eurasian *de facto* states. *Political Geography*, 63: 159–173.

Banfield, J. (2020) 'That's the way to do it!': Establishing the peculiar geographies of puppetry. *Cultural Geographies*, 28(1): 141–156. doi:10.1177/1474474020956255.

Buckingham, D. (2007) Childhood in the age of global media. *Children's Geographies*, 5: 43–54.

Connelly, F.S. (2012) *The grotesque in Western art and culture: The image at play*. Cambridge University Press: New York.

Cutler Shershow, S. (1994) "Punch and Judy" and cultural appropriation. *Cultural Studies*, 8: 527–555.

Danow, D. (1995) *The spirit of carnival: Magical realism and the grotesque*. The University Press of Kentucky: Lexington, KY.

dell'Agnese, E. & Amilhat Szary, A.L. (2015) Borderscapes: From border landscapes to border aesthetics. *Geopolitics*, 20: 4–13.

Dittmer, J. (2005) Captain America's empire: Reflections on identity, popular culture, and post-9/11 geopolitics. *Annals of the Association of American Geographers*, 95: 626–643.

Dittmer, J. (2007) "America is safe while its boys and girls believe in its creeds!" Captain America and American identity prior to World War Two. *Environment and Planning D: Society and Space*, 25: 401–423.

Dittmer, J. (2010) Comic book visualities: A methodological manifesto on geography, montage and narration. *Transactions of the Institute of British Geographers*, 35: 222–236.

Dittmer, J. & Latham, A. (2014) The rut and the gutter: Space and time in graphic narrative. *Cultural Geographies*, 22: 427–444.

Dodds, K. (2007) Steve Bull's eye: Cartoons, geopolitics and the visualization of the 'War on Terror'. *Security Dialogue*, 38: 157–177.

Duggan, R. (2016) *The grotesque in contemporary British fiction*. Manchester University Press: Manchester, NY.

Dunnett, O. (2009) Identity and geopolitics in Hergé's Adventures of Tintin. *Social and Cultural Geography*, 10: 583–598.

Gomez, C. (2014) Teaching physical geography at university with cartoons and comic strips: Motivation, construction and usage. *New Zealand Geographer*, 70: 140–145.

Goodwin, J. (2009) *Modern American grotesque: Literature and photography*. The Ohio State University Press: Columbus, OH.

Hammett, D. & Mather, C. (2011) Beyond decoding: Political cartoons in the classroom. *Journal of Geography in Higher Education*, 35: 103–119.

Harpham, G. (1976) The grotesque: First principles. *The Journal of Aesthetics and Art Criticism*, 34(4): 461–468.

Harpham, G. (2006) *On the grotesque: Strategies of contradiction in art and literature*. The Davies Group Publishers: Aurora, CO.

van Houtum, H. & Eker, M. (2015) Borderscapes: Redesigning the borderline. *Territorio*, 72: 101–108.

Katz, C. (2013) Playing with fieldwork. *Social and Cultural Geography*, 14: 762–772.

Kayser, W. (1963) *The grotesque in art and literature*. Indiana University Press: Bloomington, IN.

Kleeman, G. (2006) Not just for fun: Using cartoons to investigate geographical issues. *New Zealand Geographer*, 62: 144–151.

Laurier, E. (2014) The graphic transcript: Poaching comic book grammar for inscribing the visual, spatial and temporal aspects of action. *Geography Compass*, 8(4): 235–248.

Li, M.O. (2009) *Ambiguous bodies: Reading the grotesque in Japanese Setsuwa Tales.* Stanford University Press: Stanford, CA.

McCann, E.J. (1999) Race, protest and public space: Contextualising Lefebvre in the US city. *Antipode*, 31: 163–184.

Mello, A. (2016) Transembodiment: Embodied practice in puppet and material performance. *Performance Research*, 21(5): 49–58.

Pelletti, J.C. (1973) Once upon a continent: A puppet play for concept development. *Journal of Geography*, 72: 43–50.

Peterle, G. (2017) Comic book cartographies: A cartocentred reading of *City of Glass*, the graphic novel. *Cultural Geographies*, 24: 43–68.

Robson, M. (2019) Metaphor and irony in the constitution of the UK borders: An assessment of the 'Mac' cartoons in the Daily Mail newspaper. *Political Geography*, 71: 115–125.

Tanner, J. (2016) Geography and the creative arts. In: S. Scoffham (ed) *Teaching geography creatively*. Routledge: London, 147–162.

Thorogood, J. (2016) Satire and geopolitics: Vulgarity, ambiguity and the body grotesque in *South Park. Geopolitics*, 21: 215–235.

Thorogood, J. (2018) Cartoon controversies and geopolitics: *Archer*, animators and audiences. *Social and Cultural Geography*. doi:10.1080/14649365.2018.1497190.

# 4 Musical puppets

Associations between puppets and music are rife. Not only are puppets prevalent in musical theatre (e.g., *War Horse*, *Dr Dolittle*, *Running Wild*, *Avenue Q*), but specific puppet characters have their own musical careers and associations, and certain musical performers have their own associations with puppets. Muffin the Mule has always been at home on top of Annette Mills' piano, and The Muppets are strongly associated with music, with their films often involving numerous musical scores and individual Muppet characters playing musical instruments. Kermit's instrument of choice is the banjo; Rolf plays the piano, and Animal is on the drums, while the theme song to *The Muppet Show* – set in a theatre – clearly prioritises the musical over the visual in the theatrical as its first line is: 'It's time to play the music' (The Jim Henson Company 2002). Keith Harris and Orville reached number four in the UK singles chart in 1983 with *I Wish I Could Fly*, while the Spitting Image puppets reached number one in the UK singles chart in 1986 with their anarchic number *The Chicken Song*. More recently, Katherine Jenkins serenaded Kermit the Frog in a Christmas Special at London's Hammersmith Apollo in 2014 (Anon 2014) and Kylie Minogue performed with The Muppets at the O2 Arena in 2018 (Anon 2018). However, there are other associations between musicians and puppets, as Ed Sheeran has employed puppetry in his videos, with his puppet being exhibited in Ipswich in 2019 in advance of a series of local concerts (Anon 2019), and the Kaisers' guitarist George Miller has been making marionettes of other musicians (Anon 2020).

By contrast, geography is often characterised as an explicitly visual discipline, with an emphasis on mapping practices and a predominance of research targeted towards written texts and visual imagery. This emphasis has been critiqued by numerous geographers of sound and music (e.g., Kong 1995a; Smith 1997; Back 2003; Duffy 2005, 2009), and the geographies of music have experienced considerable growth over recent decades (Smith 1994; Kong 1995a; Keeling 2011). Compounding the historic prioritisation of the visual over the aural, leading to an analytical emphasis on lyrics and scores rather than practice and performance (Frith 1996; Duffy 2005; Wood 2012), has been the elitist focus on supposedly higher artforms over lower artforms, with popular music being especially marginalised as base entertainment compared to more esteemed classical music (Leyshon et al. 1995; Kong 1995a).

DOI: 10.4324/9781003214861-10

Recent developments place more emphasis on the doings of music, and within this diversity of work are two primary relationships – that between music and place, and that between music and identity – with each heavily implicated in the other. For example, the role of music in actively producing space – private, public, lesbian and so on – and specific places – through associations with individual performers, groups and genres – has been variously explored (see, for example, Valentine 1995; Kruse II 2005; Daynes 2009; Keeling 2011; Born 2013). Similarly, the forging and contestation of national identity have been examined in the context of intersections between hegemonic and counter-hegemonic narratives (Kong 1995b) and between tradition and novelty (Wood 2012), firmly embedding music, place and identity in each other. Music, then, is now conceived not as a thing but a practice, which is generative of identities (of people and of places) in the intersection of multiple and diverse bodies, objects and relations (Frith 1996; Small 1998; Anderson et al. 2005; Duffy 2005; Wood et al. 2007; Keeling 2011; Revill 2015). Such work also acknowledges that meaning is as much about affectivity as intellectualism (Wood et al. 2007; Wood 2012), paving the way for greater engagement with popular music and incorporation of everyday, lay musical experiences rather than confining musical analysis to those formally trained (Frith 1996; Elliott 2000; Smith 2000; Wood et al. 2007).

With the geographies of music now focusing less on what music is and more on what music does (Small 1998; Daynes 2009; Born 2013), and clear calls for greater attention to be paid to everyday musical experiences and articulations (Wood et al. 2007), it is acknowledged that the musically untrained might still be able to contribute to musical analysis based on their quotidian experiences of music, even if that contribution is articulated in lay terms rather than technical language. It is in this democratic spirit that I undertook the analysis presented below, as I am not musically trained (although I dabbled with recorder and alto saxophone in my school days) and I do not consider myself to 'be musical'. I make no attempt to analyse musical scores, nor am I equipped to undertake formalist analysis of compositional features. However, I have grown up as surrounded by music as the average person (e.g., school assemblies, pop music, advertising, film scores, muzak, musical theatre and so on) and I am aware both of my own musical preferences and the ways in which music affects me in cognitive, bodily and affective ways. Consequently, my emphasis is on my role and experience as a consumer of music, focusing on practices and experiences of listening, supplemented by lyrical analysis where relevant.

To this end, I take seriously the claim that it is legitimate to listen for ways in which musical patterns describe, represent or convey people, places or ideas and I set out to listen attentively and actively (Elliott 2000; Smith 2000). I am concerned less with documenting what those musical patterns are than I am with capturing what they do, in cognitive, bodily and affective ways, that contributes to musical constructions of puppets. In this vein, I align my work in this chapter with Walser's (2003) advocacy of a self-reflective and tactical analysis that interprets rather than labels music, that is founded on the discursive competency of the analyst and that should be evaluated according to its aims. Those aims are to establish how musical works relating to puppets construct the idea of the puppet and how this varies between the notion of 'the puppet', specific types of puppet and individual puppet characters.

In total, I examined works by 15 composers or performers, which were identified through personal knowledge and internet searches and for which I could readily source the relevant work/s. While the titular emphasis of this volume is on popular culture, I am mindful of the problematic distinction between popular and classical music, and – in any event – I could not source much popular music that related to puppets, so this analysis considers both music that we would conventionally think of as classical and the music that falls outside of that category, which I term here 'modern' as a catch-all label, while also identifying musical style in more detail within those broad categories. The works are also grouped according to whether they relate to puppets in a general sense, to a certain type of puppet (e.g., marionettes) or to a specific and identifiable puppet character (e.g., Mr Punch). Within the 'modern' musical works, a further distinction is drawn between works that are about puppets and works that are performed by puppets. Table 4.1 specifies the puppet-relevant works that I could both identify and source for analysis, and the details on both how they relate to puppets and how they compare with each other as musical works.

Four caveats are worth noting with respect to this classification. Firstly, *The Wooden Prince* is included despite this character being described as a doll rather than a puppet in the CD sleeve because it has also been described as a puppet, both in published works (see, for example, Tazudeen 2020) and in the CD insert (Palmer 1991). By contrast, Delibes' *Coppelia* has not been included as the animated dolls in this work are explicitly mechanical dolls (they are wound up) rather than puppets. Secondly, the selected tracks for Martinů and Macdowell were chosen either because they are relevant to the idea of puppets and marionettes respectively in an explicit but general sense rather than relating to non-puppet cultural forms (such as Harlequin) or because they puppetise different affective states (e.g., shyness) or contrasting caricatures (e.g., witch, clown), which collectively could inform their constructions of puppets more generally. Thirdly, the two works by Stravinsky – *Petrushka* and *Pulcinella* – relate to different cultural forms (Russian and Italian respectively) of the same anarchic trickster (who is akin to Mr Punch in his anglicised form), although the character in the ballet is Pierrot rather than the Russian puppet practice and form of Petrushka (Taruskin 1996). Fourthly, and finally, *The Puppet Song*, by TryHardNinja is associated with an online game (*Five Nights at Freddy's*) and is sung by an animated character called The Puppet, although the video emphasises the animated rather than the puppet nature of this character.

Through these selections, an array of works by diverse composers and performers was identified, with three composers/performers within each sub-group to provide consistency in the comparative basis for the analysis. My analytical process involved three stages. Firstly, I listened attentively to each

*Table 4.1* List of works analysed

| Categories | | Name | Nationality | Title | Year | Style | Work type | Length |
|---|---|---|---|---|---|---|---|---|
| Puppets | Classical | Debussy | French | Fêtes Galantes Livre 1 no 2: Fantoches | 1904 | Voice and piano | Single track | 1:25 |
| | | Martinů | Czech | Loutky (Puppets), I, II, III | 1914–1924 | Solo piano | Selected tracks: New puppet Shy puppet Sick puppet | 1:33 3:25 3:43 |
| | | Bartók | Hungarian | The Wooden Prince | 1914–1916 | Ballet for orchestra | Symphony | 54:04 |
| | Modern, about puppets | Sandie Shaw | British | Puppet on a String | 1967 | Pop | Single track | 2:23 |
| | | Tyler, the Creator | American | Puppet | 2019 | Rap | Single track | 2:59 |
| | | Metallica | American | Master of Puppets | 1986 | Heavy metal | Single track | 8:35 |
| | Modern, by puppets | Keith Harris and Orville | British | I Wish I could Fly | 1982 | Pop | Single track | 3:12 |
| | | Spitting Image | British | The Chicken Song | 1986 | Pop | Single track | 2:36 |
| | | TryHardNinja | American | The Puppet Song | 2015 | Online game song | Single track | 3:54 |

| | | Composer | Nationality | Work | Year | Format | Tracks | Duration |
|---|---|---|---|---|---|---|---|---|
| Marionettes | Classical | Gounod | French | Funeral March of a Marionette | 1872/1879 | Solo piano, orchestra | Single track | 4:36 |
| | | Macdowell | American | Marionettes Op 38 | 1888/1901 | Solo piano | Selected tracks: Prologue Witch Clown Epilogue | 1:16 0:48 2:14 1:26 |
| | | Cui | Russian | Miniatures Op 39 no 1: Marionettes Espagnoles | 1886 | Solo piano | Single track | 1:01 |
| | | Goossens | British | Kaleidoscope, 18, 6: Punch and Judy Show | 1917 | Solo piano | Single track | 0:42 |
| Punch | Classical | Stravinsky | Russian | Petrushka | 1911 | Ballet for orchestra | Symphony | 36:24 |
| | | Stravinsky | Russian | Pulcinella (Suite) | 1920 | Ballet for orchestra | Symphony | 25:29 |
| | | Birtwistle | British | Punch and Judy | 1967 | Chamber opera | Symphony | 1:39:24 |

piece at least twice and noted down thoughts, associations, senses, interpretations and feelings that they evoked. Subsequently, I reflected upon those notes as a single collection and within their sub-groups (puppet, marionette, Mr Punch) in the context of the research questions. Finally, I engaged with formal reviews of a selection of the works, ensuring that there was one such work in each sub-group, to consider my own responses and interpretations in the context of more informed musical perspectives, listening again to any works that I wanted to explore further.

In the discussion that follows, the works analysed are variously identified by name of composer/performer or by title, depending on the discursive context. I begin by considering representations of puppets in the lyrics analysed, which identifies similarities to the metaphorical understandings of puppets discussed in Part 1. Subsequently, I explore my responses to the music within the whole sample, which is oriented around mood and movement on the one hand, and contrasts and tensions on the other hand. Finally, I develop direct responses to the two research aims and set these within my conceptual concerns for the topological gulf, body grotesque and borderscaping.

## Lyrical puppets

Of the six modern works analysed, three are about puppets and three are nominally sung by puppets. Each group is addressed in turn, drawing out similarities and differences within and between the groups (e.g., theme, tone and affect), relating these back to constructions of puppets identified in the preceding chapters and considering intertextual influences on the appreciation of these works.

### Songs about puppets

Attending first to song lyrics referring to puppets, one (*Metallica*) is a cautionary tale about drug addiction and two (*Sandie Shaw*; *Tyler, the Creator*) are love songs. These references to puppets are metaphorical and several aspects of these works reinforce the analytical findings of Part 1, most evidently in relation to assumptions of a specific type of puppet (marionettes) and a lack of agency or power on the part of the person being puppetised. The titular reference to string in Shaw's 1967 winning entry to the Eurovision Song Context clearly establishes this puppet as a marionette, and this stringy emphasis also shows through in *Metallica*'s Master of Puppets, with the phrase 'I'm pulling your strings' being a recurrent refrain. However, there is no mention of faciality, in contrast to the focus on ventriloquist's dummies in passing literary metaphorical references to puppets (Chapter 1). While strings suggest that the character is being controlled on a bodily level, all three songs about puppets emphasise strong psychological influences, yet each work indicates a different level of awareness of this influence. *Shaw*, for example, is clearly aware of the dependence of her own actions on the declaration of love by the target of her affection but *Tyler, the Creator* both indicates a gradual realisation of the extent of dependence on his loved one ('I'm starting to wonder') and sings of a loss of self-knowledge resulting from this dependence ('I don't know me'). However, for *Tyler, the Creator*, this appears to be a temporary state of self-ignorance, as the concluding line evokes a sudden and potentially unpredictable reawakening ('But at some point, you come to your senses'). By contrast, the controlling power of drugs in *Metallica* is more totalising. In this instance, there is an eventual and inevitable loss of perception, not only of the self but of anything, and there is no association drawn between succumbing to that control and the attainment of a desired outcome, as could be discerned in the two love songs. In a graphic description of the psychotropic power of drugs ('Twisting your mind and smashing your dreams'), the destruction of the person, their capacities and their future is made clear.

The different tone in the lyrics of each work is also notable. Relentlessness shows through clearly in *Metallica*'s lyrics, with extensive repetition of individual words, phrases and refrains, giving a strong sense of being compelled towards an inevitable demise, but it is the contrast in tone between *Shaw* and *Tyler, the Creator* that is most interesting in this regard. The tone of *Shaw's* lyrics is at most curious as to what the future holds ('But where's it leading me to?') even though she articulates awareness of her vulnerability to the whims of her loved one. By contrast, the tone of *Tyler, the Creator's* lyrics is tense and angry, expressing confusion ('Do you need a hug? Do you need to be alone?') alongside frustration and resentment. *Shaw* remains hopeful of a happy romantic outcome but *Tyler, the Creator* is worn down and frustrated both with the person whose company he craves and with himself for his failure to break free of the situation he is in. At the end, however, *Tyler, the Creator* finds release from this scenario, starting to speak in the past tense ('I've been lost') and acknowledging an ultimate return to self-awareness and agency, coming to his senses in the final line. The emotional fervour, then, is distinctly different in these two works: The negative intensity in *Tyler, the Creator's* Puppet is ultimately resolved despite not succeeding in its quest, while *Shaw* is left permanently hopeful yet unfulfilled.

Collectively, these works conjure a puppet that is a marionette, which lacks a face and which is highly psychologised, in a different form of body grotesque to that identified in Part 1 that is at the same time a character contorted, as evidenced by the romantic dependence on another and different degrees of self-awareness as to their dependency. However, the physical form of these metaphorical puppets is not equally specified in the three works. Whereas both *Shaw's* and *Metallica's* references to string evoke an image of a marionette, *Tyler, the Creator* gives no indication of the form of the puppet referred to, aligning the first two more closely with passing literary metaphorical references to puppets and *Tyler, the Creator's* puppet more closely with narrative literary metaphorical uses. These works, then, in their mixture of similarities of features (lack of self-control, lack of faciality), fit well with the earlier analyses of metaphorical uses of puppets. However, they also generate very different affective qualities and impacts (e.g., with respect to tone and outcome), and reflect, to different degrees, the passing and narrative metaphorical uses of puppets found in popular fiction, while simultaneously reconfiguring a composite metaphorical puppet as a faceless and strongly psychologised marionette in a distinct formulation of body grotesque and character contorted.

### Songs by puppets

Progressing now to explore songs produced by puppets, the lyrics to *The Chicken Song* are essentially nonsense, but the bonkers ideas promoted in the song are entirely in keeping with the 'anything is possible' nature of puppets. While burying all your clothes or painting your left knee green is an entirely feasible action for humans, puppets – unlike humans – would be able to eat a Renault Four with salami in their ears and disembowel themselves with spears, if they chose to do so. While puppets here are used as proxies for human performers and stand as representations of identifiable public figures from the time, they are also able to articulate ideas that would be highly politically incorrect (and illegal) – 'casserole your gran' – in a reflection of both the social mediation afforded by puppets that was identified in Chapter 3, and the puppet's potential for chaos and waywardness. Moreover, the essence of the song aptly reflects the contrariness of puppets. The song freely admits that it is nauseating, loud and grating to the extent that people will hate it, but the song is also proclaimed to be inescapable as it has the quality of an ear-worm, mentally replaying itself over and over despite being unbidden and unwanted ('And though you hate this song, You'll be singing it for weeks').

This song, then, evokes a rowdy, party-like atmosphere of bizarre behavioural opportunities that – while not about puppets – reflects various qualities of puppets, whether intentionally or incidentally. It would be difficult for any song to be more different to this than *I Wish I Could Fly*. This is a melodic duet with lullaby intonations that narrates an emotional dilemma on the part of Orville (a green fluffy duckling puppet who sports a nappy) and the nature and strength of the relationship between the duck and his side-kick, Keith Harris. While also not being about puppets, this song is equally reflective of certain qualities of puppets already identified.

Orville is lacking in self-esteem and is depressed, which is established through his articulated wishes to be able to fly and to have parents, so Orville is a feeling and sensitive rather than an anarchic or dehumanised puppet. Keith Harris provides a counter-narrative to try to convince Orville that he can fly, that he is not alone, that his sadness will end and that he is not daft. While the lyrics establish a clear and strong relationship between the puppet and puppeteer, the nature of this relationship is somewhat ambiguous as in the first and penultimate verses Keith Harris positions himself as Orville's 'very best friend' but in-between Orville asks Keith Harris if he will be there to tuck him in at night, suggesting more of a parental relationship consistent with Orville's desire to have parents.

One striking aspect of the lyrics is the permanence of the relational ties between the two characters. In verses one and four, Keith Harris says that 'I know we'll never part' and in verse three undertakes to 'always be there' for Orville. However, it is also eerily reminiscent of the permanence of the person–puppet relationship identified in common puppet tropes in horror films. Although this is partially offset by the saccharine sweetness of both the lyrics and melody, this sweetness itself takes on a different hue when read in the context of previous findings about the strategies through which puppet masters cultivate a relationship of dependency between themselves and the target of their control. The repeated assertion by Keith Harris to be there to help the forlorn Orville could be interpreted as fostering dependence, with 'I'm gonna help you mend your broken heart' being proffered four times. Similarly, the repeated questioning by Keith Harris as to who Orville's very best friend is, puts Orville in the position of being expected to say 'you are', which he duly says, three times. Through behaviour that can easily be construed as coercive, Orville becomes embedded within the relationship that Keith Harris perceives or desires them to have. Despite the tweeness of the music and lyrics, the lyrics can become very uncomfortable when set in the context of other cultural uses of puppets. On this intertextual reading, the lyrics powerfully reflect the contradictoriness of seemingly benign behaviours that mask pernicious exploitation, raising questions as to whether the song really is saccharine sweet or suspiciously sinister.

Finally, *TryHardNinja's* song originated in an online game and the titular puppet is possessed by the spirit of a dead girl, reinforcing the link identified in the previous chapter between puppets and horror. There are also allusions to marionettes as a specific type of puppet, through reference to cutting the strings as a means of escape but this is magnified by the subsequent description of walking in chains, thereby emphasising lack of agency. The Puppet describes itself as powerless and as wearing a mask that it is trapped inside, so although faciality does feature, this is a hidden or unknown faciality and it is not self-knowledge that is lacking but recognition by others. Indeed, The Puppet is very aware of its predicament, and undertakes to shelter and comfort the other characters in the game ('I'm powerless to change your fates, But in the end I'll shelter you my friends'). While acknowledging its own powerlessness to change their predicament, The Puppet expresses agency within the context of that predicament, introducing a nuanced perspective on the lack of control associated with puppets. By putting itself in a position of superiority or maternalism with respect to the other characters in the game, The Puppet effectively takes Keith Harris's place, exerting a purportedly protective relationality towards the other characters, who are effectively puppetised.

*TryHardNinja*, then, both reinscribes several of the features associated with puppets in previous analyses, but also injects different perspectives, specifically in relation to issues of faciality, recognition by self or others and lack of control. With respect to the last of these, despite the seemingly benign intentions, the possibility remains that this is deceptive, consistent with the narrative metaphorical uses of puppets in fiction and bringing us back to the contrariness of the puppet. If the saccharine sweetness of Keith Harris and Orville can be interpreted in a much more insidious sense, then one can only imagine what might lie ahead for the characters that have been puppetised by The Puppet in the horror setting of *Five Nights at Freddy's*.

## Musical puppets

Looking back over the notes of my analysis, two distinct organising themes emerged: The first concerns mood, movement and how they relate to each other; the second revolves around contrasts or tensions and leads to consideration of progressions. However, these two thematic concerns are not entirely discrete, and these intersections between mood/movement and contrasts/tensions provide their own insights into the musical and broader cultural construction of puppets, enabling further development of both our understanding of the contrariness of the puppet and the varied configurations of body grotesque and character contorted that constitute puppets.

### Mood and movement

A panoply of moods was detectable across the works, consistent with the affective ebbs and flows of the narratives being told in the longer works and the range of human emotions and experiences explored in the shorter works. Moods that I noted included infectious gaiety (e.g., *Pulcinella*; *Marionettes Espagnoles*), heaviness of heart (e.g., *Clown*, *The Wooden Prince*), contentedness (e.g., *Pulcinella*), apprehension (e.g., *Petrushka*), aggression (e.g., *Punch and Judy*; *Petrushka*), melancholy (e.g., *Petrushka*, *The Puppet Song*) and affection building into sexual tension (*Pulcinella*; *The Wooden Prince*). However, given the interchangeability of human and puppet characters in many of these works – whether in the form of puppets coming to life in human form (e.g., Stravinsky's *Petrushka*; Bartók's *The Wooden Prince*), the exploration of human traits through puppets (e.g., Martinů's *shy* and *sick puppets*) or the description of humans as puppets (e.g., *Puppet on a String*; *Puppet*) – it is difficult to associate any particular mood with either puppets (as distinct from humans) in general or with specific types of puppets.

Movement could more readily be associated with puppets as distinct from humans, consistent with puppetry literature that emphasises the significance of rhythm and finding movement in the form of the puppet for puppetry to be effective (Schechner 2002; Francis 2012). This was especially so in those works which are either explicitly about puppets or which entail puppet–human transformation. Musical works explicitly and solely about puppets are useful because it is just the puppet rather than – for example – a mood that is being evoked. Works entailing puppets becoming human are helpful because they provide a direct contrast between the composer's evocation of human versus puppet motor capabilities and characteristics. In these contexts, two types of movement emerged as distinguishing puppets from humans, which I begin by characterising here in colloquial terms – drawing directly from my notes – as 'plinky-plonky' and 'perky-jerky'. As a starting point, plinky-plonky refers to a lightness and exaggeratedness of movement and perky-jerky refers to an uncoordinated and disjointed motion, but these terms are developed into something more conceptually rigorous through the paragraphs below.

Plinky-plonkiness is aptly evidenced in *Puppet on a String*, with its upbeat, jingly, over-the-top roller-coaster of musical expression that conveys very effectively the affective ups and downs of the imagined future in the reality of the present, but similar jingly intonations show through in other works, most notably *Fantoches*. However, plinky-plonkiness is less an absolute quality that a work either does or does not possess and more a continuum, with ·*Puppet on a String* at the high end and *Funeral March of a Marionette* and *Marionettes Espagnoles* at the low end. In some cases, I associated these musical expressions with oversized vertical leg movements in creating a step or with an excessive arcing reach of the arms, but my sense of plinky-plonk-iness is less about the scale of movement than the spirit of movement, char-acterised by jauntiness rather than fluidity, but not to the extent that such movements become jarring or incoherent. The 'plinky' encapsulates the light-ness, and the 'plonky' speaks to a sense of self-placement that is not entirely controlled but is co-ordinated, in which contact with the ground is as exag-gerated as the steps and sweeps of the legs and arms. In more formal par-lance, then, my scribbled notes of 'plinky-plonky' music could be described in terms of a grounded lightness generated by exaggerated jauntiness.

Perky-jerkiness, on the other hand, is more about a lack of co-ordination, a denial or repression of fluidity, which evokes disjointedness, incoherence and even disaggregation of body parts, as evident in *Punch and Judy's* stark contrasts between very clipped and very protracted vocal enunciations and its (at times) seemingly unstructured musical erraticism. However, this too should be considered a continuum, with its other extreme characterised by the deadness of movement, as apparent in *The Wooden Prince*, when the Princess struggles to get the Wooden Prince to dance, and the musical expres-sion shifts from plinky-plonky self-placement to effortful thrust and shove. With perky-jerkiness, the 'perky' evokes spontaneous uncontrollability and lack of co-ordination, and the 'jerky' refers to the opposite to this extreme motoric erraticism, as extreme material auto-immobility. A perky puppet might voluntarily exhibit seemingly uncoordinated control of its own limbs, but a jerky puppet can only move due to external forces and only as a singular and lumpen form. In more formal parlance, then, perky-jerkiness might be described in terms of an anatomical incoherence, which takes extreme forms in either the seeming disassembly of a previously cohesive anatomy or the solidification of a previously cohesive anatomy, such that it is still cohesive as a singularity but no longer has meaningful anatomy or mobility.

Clearly, these two continua can interact, as the exaggerated jauntiness of the plinky-plonky can be articulated musically as either self-directed and therefore co-ordinated or as externally enforced as the extreme erraticism of the perky-jerky. However, they are also in some ways mutually exclusive, as the deadness of the materially immobile singularity cannot exhibit exaggerated jauntiness as it has no distinct anatomy with which to be jaunty and no motoric capacity to do so, either. These varied musical evocations of puppets, then, culminate – for me – in two distinct expressions of puppet movement: Grounded lightness generated by exaggerated jauntiness and anatomical disassembly, either through disaggregating different body parts or doing away with the body as an assembly of parts. Consequently, while mood and movement can be considered discretely in the musical construction of puppets, they are also held in creative tension, with an ambiguous and malleable border between them, amenable to diverse modes of borderscaping.

### Contrasts and tensions

Supplementing this creative tension, a host of additional contrasts and tensions can be discerned in these works, all of which feed into the musical construction of puppets. *Petrushka*, for example, combines musical evocation of humour with seriousness and lightness with sadness, *Metallica* exhibits both urgency and constancy, and *Tyler, the Creator* integrates serene melody with abrupt agitation. Similarly, *Gounod* intermixes gentility with abrasiveness, and *Stravinsky* combines both routine and event (*Pulcinella*) and the festive and the ominous (*Petrushka*). In some cases, these contrasts ebb and flow through the work, for example in the context of the narrative in longer works, as plot, characterisation and affective qualities shift (e.g., *Petrushka*, *Pulcinella*), but in other cases, the oppositional qualities are held in tension (e.g., *Puppet*). While such contrasting musical evocations are clearly consistent with the contrariness of the puppet, they also suggest that although contrariness is a characteristic feature of puppets, the simple existence of contradictions is not sufficient to establish a topological gulf, as for such a gulf to exist the distinct elements that it holds apart must be contemporaneous: They must be held in tension if the border between them is to be theatrically dissolved. Otherwise, one simply replaces the other. In addition, they shine an ontological spotlight on both humans and puppets and further ambiguate the relationship between mood and movement.

In terms of the human, *Martinů* explores and evokes not only the human experience of being new, shy or sick but also the progression of the human through shyness and sickness. In *New Puppet*, there is a sense of increasing confidence on the part of the puppet in its own abilities as the musical expression proceeds from stepping to dancing to leaping, with the puppet becoming progressively emboldened, gregarious and flamboyant. In *Shy Puppet*, the trepidatious puppet battles with its own timidity, initially expressing wariness and caution before rushing anxiously in a release of nervous energy and then finding a sense of mastery in the face of an uncertain environment. Finally, in *Sick Puppet*, the poorly puppet begins slowly, devoid of energy, but becomes intermittently more upbeat in a gradual, staged recovery of strength and mood, which features periodic relapses, returning to sombre quietude in advance of another period of improving health. However, the relationship between mood and movement varied between them, at least in my experience. With *New Puppet* and *Shy Puppet*, it was the nature and pace of the music that described the puppet's movement for me, from which I then extrapolated an understanding of their mood, motivations, perceptions and actions, but with *Sick Puppet*, it was the mood that dominated over the movement. While the first word that I jotted down was 'slow', the second word was 'melancholy'. 'Upbeat', 'cheery' and 'brightening of mood' also feature in my brief notes (consisting of only six lines), and these notes explicitly acknowledge that the puppet seems more sad than sick. Consequently, while the music can suggest either the mood or the movement as being the determining aspect of a construction of a puppet, mood and movement can relate in ambiguous ways, which is compounded by the strategic and aesthetic manipulation of grounded lightness and anatomical disassembly through which these moods and movements are established.

Moreover, the musical expressions of mood and/or movement are not necessarily consistent with the 'type' of puppet seemingly being constructed. In MacDowell's series of works on marionettes, the witch and the clown stand out as examples of puppet contrariness, as the mood of each runs counter to that which I expected. The title *Witch* suggests evil, cruel magic, but the music to me seemed childlike and light-hearted, appropriate for a nursery, and suggestive of a benevolent, white witch in stark contrast to the source of childhood frights that I was expecting. Conversely, *Clown* might be expected to be fast-paced, comedic and jocular, but I found this music to be considered, sedate, melancholy even, more appropriate for a wake than for the Big Top. While these examples clearly reflect the contrariness of the puppet, they also remind us of additional aspects of puppets. Gounod's *Funeral March of a Marionette* is similarly contradictory in its musical evocation compared to the thematic emphasis of the work and associated affective expectations, as it did not seem to me to evoke a funeral. It seemed too energetic, too lilting and light-hearted. While darker undercurrents were discernible, it reminded me more of a teddy bears' picnic than a marionette's funeral. Then again, that perhaps is precisely the point, drawing attention to the fact that puppets cannot die and highlighting their capacity to mock the human condition even while mimicking it. Yet even acknowledging the contrariness of the puppet does not override the contradiction between the expectation and the experience, sustaining the constitutive tension between the two as another topological gulf.

Perhaps most tellingly, though, is the musical depiction of Mr Punch in Birtwistle's *Punch and Judy*. Further evidencing the contrariness of puppets, Punch himself embodies contradictory characteristics of paternal tenderness and sociopathic viciousness (shown most clearly in the speed with which he switches from cradling the baby to throwing it onto the fire) and the play incorporates narratives of both romantic love (for Polly) and murderous brutality (towards the baby, Judy, the doctor and lawyer, to name a few). While these are firmly established in both the lyrics and music of the opera, my overall sense of this work is that it worked less effectively in characterising Punch as a distinct character and much more effectively in characterising

Punch as the extreme archetype of a puppet. Among Punch's many distinguishing features (e.g., humpback, hooked nose, jingly hat, slapstick), the most easily identifiable is his squeaky, squawky and shrill voice. There is plenty of squeaky, squawky shrillness in the *Punch and Judy* opera, but not in the vocalisation of Punch, whose voice I described as 'deep, measured and slow' and at times was difficult to differentiate from that of another character. Similarly, while there was bountiful discordance and abrasiveness in the music and vocals, the freneticism with which I associate the Punch and Judy puppet show seemed somewhat lacking, especially – again – from Punch himself. The pace and tone of Punch in both words and music seemed more furtive than frantic, more watchful and opportunist than cunning and devious, crudely and habitually violent rather than tactically deceptive and coercive; so much so that he was more reminiscent to me of the Child Catcher from *Chitty Chitty Bang Bang* than of the titular Punch.

For me, this is a clear case of the character contorted, but in this instance, it is not the human character that is contorted, but the puppet's character. Both the vocal and energetic characterisation of Punch were constituted more in the percussive background than in the figural foreground. However, this discordant, abrasive, chaotic and frantic musical and choral context very effectively evoked both the mismatch between Punch's amorous pursuit of Polly and his cold-hearted bludgeoning of everybody else, and between Punch's character, actions and aspirations on the one hand and the behavioural and moral norms of society on the other. This musical background not only reinscribes the contrariness of puppets and situates Punch firmly within this contrariness, but also serves to disembody the character of Punch from the operatic figure of Punch by translocating some of Punch's defining characteristics from Punch himself to the milieu within which he is performed into being. He becomes the epitome of the contrariness of the puppet, in relation to both the Janus-faced nature of his own actions and the incongruence of his actions with respect to social norms, but while his own status as puppet is reinforced in this networked redistribution of character, his uniqueness as Mr Punch is diluted, as this uniqueness has been shifted from the figure to the background. In effect, Punch is de-Punchified as a distinct character while simultaneously being hyper-puppetised through this de-Punchification. This form of the character contorted is reminiscent of the evacuation of the subject in Part 1, as the subjectivity of Punch is not eliminated entirely but is redistributed to the musical and choral context within which he resides. It also enables us to tease out two further topological gulfs from this example – that between archetype and character and that between figure and ground – that (in their own way) both sustain and eliminate (theatrically dissolve) the border between their respective elements in reconfiguring Mr Punch.

Ultimately, then, this discussion of musical puppets has generated another perspective on both the body grotesque and the character contorted. From the discussion of mood and movement, we gained an understanding of the second continuum as a bodily reconfiguration through either the disassembly of anatomical body parts or the disavowal of anatomical body parts, each of which generates its own distinct movement potentialities and each of which instantiates its own body grotesque. From this latter discussion of contrasts and tensions, we reached a new understanding of what can be meant and generated by the character contorted, especially in relation to the detailed consideration of the construction of Mr Punch in Birtwistle's *Punch and Judy*, which highlighted the relational de/construction of this puppet. By transferring some of his defining features to the musical and choral background, his puppetness (contrariness) is universalised, he is hyper-puppetised as the epitome of this contrariness and he is de-Punchified or diluted as a unique puppet character.

## Conclusion

Through a focused analysis of diverse musical works that relate to puppets in one form or another, this chapter both reinforces findings and perspectives that emerged in the preceding chapters and provides new insights into and understandings of the construction of puppets in popular culture. Despite the naïve or lay nature of the analysis undertaken, this discussion also enables us to answer the two research aims that guided the musical analysis of puppets. Addressing first the question of how musical works relating to puppets construct the idea of the puppet; this can be considered from both lyrical and musical perspectives.

In lyrical terms, puppets are constructed predominantly in 'modern' musical works, and as a composite form, the puppet here is a faceless marionette that is highly psychologised as the puppetised character's capacity for self-directed action and even their self-awareness is diminished through the cultivation of a relation of dependency. Some variation on this was evident, as *Puppet* is not suggestive of a marionette and *TryHardNinja* injects nuance into the loss of agency through the potential puppetisation of other characters by The Puppet of the title. However, the parallels between the constructions of puppets in these modern musical works and those identified in Part 1 are clear, albeit that the musical puppet takes slightly reworked form as the body grotesque of this puppet is faceless (although this is itself a distinct faciality) and the musical puppet integrates rather than isolates the body grotesque and the character contorted.

In musical terms, puppets are constructed through both the evocation of mood and movement and through the employment of contrasts and tensions. An explicitly puppet mood was difficult to discern but puppet movement was found to be articulated musically through two continua. The first of these – 'plinky-plonky' – was developed into the notion of grounded lightness generated through exaggerated jauntiness in which a bouncy quality of movement is paired with a weightiness of tread or contact. The second continuum – 'perky-jerky' – was developed into the notion of anatomical disassembly, characterised by a lack of co-ordination and exhibiting two extreme forms: The seeming disaggregation of a formerly cohesive body of parts and the disavowal of the possession of anatomical parts.

This creative tension formed the link to the examination of contrasts and tensions in the musical construction of puppets, which were found to typify and reinscribe the contrariness of puppets and drew attention to the potential for musical constructions of puppets to contradict the expectations of the listener as a new topological gulf. Most tellingly, Birtwistle's *Punch and Judy* exhibited both contrasts between some of Mr Punch's key characteristics and how they are conveyed in the opera and tensions between Mr Punch's characterisation as figure and as ground, with the latter being accompanied by a tension between archetype and character as further topological gulfs. This work very creatively hyper-puppetised Mr Punch as the epitome of the contrariness of puppets while simultaneously diluting Mr Punch as a unique puppet character by translocating core definitional features from the puppet character to its musical and choral context in a striking double manoeuvre of constructing a distributed body grotesque that incorporates both figure and ground and both a depersonalised and universalised contortion of the character of Mr Punch.

Considering next the question of how musical constructions of puppets vary between puppets in general, specific types of puppet and individual puppet characters, few clear differences emerge and indeed, the different categories of puppet seem to blur into one another. In the lyrics from modern musical works about puppets in general, the emphasis was predominantly (although not exclusively) on marionettes, while the grounded lightness generated through exaggerated jauntiness is also more readily associated with the string-based control of a marionette compared to – for example – either a ventriloquist's dummy (which tends to be immobile) or a glove or sleeve puppet (especially if this is performed in relation to a performance platform as the movement would tend to be horizontal rather than vertical). Most significant, again, though, is Birtwistle's *Punch and Judy* and the way in which it both generalises Mr Punch's character by distributing it from the figure to the ground and hyper-puppetises Mr Punch as the extreme form of a generalised puppet: The epitome of contrariness.

Overall, then, while there are clearly discernible – even if sometimes inelegantly expressed – means by which puppets are constructed in musical works, there are also identifiable tendencies both to foreground the marionette and to blend different types of puppet and puppet characters together. The borders between different types of puppets are drawn diversely and flexibly through their musical construction, with topological gulfs being variously widened or bridged, in an eclectic mix of borderscaping practices that generate unique configurations of body grotesque and character contorted.

## References

Anderson, B., Morton, F. & Revill, G. (2005) Practices of music and sound. *Social and Cultural Geography*, 6(5): 639–644.

Anon. (2014) Frog chorus: Jenkins goes green. *The Daily Telegraph*, 24 December 2014, n.p.

Anon. (2018) She should be so lucky. *The Daily Telegraph*, 15 July 2018, n.p.

Anon. (2019) Unseen Ed. *The Daily Telegraph*, 16 August 2019, n.p.

Anon. (2020) Star turn. *The Daily Telegraph*, 01 August 2020, n.p.

Back, L. (2003) Deep listening: Researching music and the cartographies of sound. In: M. Ogborn, A. Blunt, P. Gruffudd, D. Pinder & J. May (eds) *Cultural geography in practice*. Routledge: Milton Park, 272–285.

Born, G. (2013) *Music, sound and space: Transformations of public and private experience*. Cambridge University Press: Cambridge.

Daynes, S. (2009) A lesson of geography, on the Riddim: The symbolic topography of Reggae music. In: T.L. Bell & O. Johansson (eds) *Sound, society and the geography of popular music*. Routledge: Milton Park, 90–102.

Duffy, M. (2005) Performing identity within a multicultural framework. *Social and Cultural Geography*, 6(5): 677–692.

Duffy, M. (2009) Methods: Sound and music. In: R. Kitchin & N. Thrift (eds) *International encyclopedia of human geography*. Elsevier, Amsterdam, 230–235.

Elliott, D.J. (2000) Music and affect: The praxial view. *Philosophy of Music Education Review*, 8(2) 79–88.

Francis, P. (2012) *Puppetry: A reader in theatre practice*. Palgrave Macmillan: Basingstoke.

Frith, S. (1996) Music and identity. In: S. Hall & P. du Gay (eds) *Questions of cultural identity*. SAGE: Los Angeles, CA, 108–127.

Keeling, D.J. (2011) Iconic landscapes: The lyrical links of songs and cities. *Focus on Geography*, 54(4): 113–125.

Kong, L. (1995a) Popular music in geographical analyses. *Progress in Human Geography*, 19(2): 183–198.

Kong, L. (1995b) Music and cultural politics: Ideology and resistance in Singapore. *Transactions of the Institute of British Geographers*, 20(4): 447–459.

Kruse II, R.J. (2005) The Beatles as place makers: Narrated landscapes in Liverpool, England. *Journal of Cultural Geography*, 22(2): 87–114.

Leyshon, A., Matless, D. & Revill, G. (1995) The place of music: Introduction. *Transactions of the Institute of British Geographers*, 20(4): 423–433.

Palmer, C. (1991) Untitled introduction to The Wooden Prince. In: J. Ginn (ed) *Bartók, The Wooden Prince: Complete ballet*. Chandos Records: Colchester, 4–7.

Revill, G. (2015) How is space made in sound? Spatial mediation, critical phenomenology and the political agency of sound. *Progress in Human Geography*, 40(2): 240–256.

Schechner, R. (2002) Julie Taymor. From Jacques Lecoq to The Lion king: An interview. In: J. Schechter (ed) *Popular theatre: A sourcebook*. Routledge: London, 64–77.

Small, C. (1998) *Musicking: The meanings of performing and listening*. University Press of New England: Hanover.

Smith, S.J. (1994) Soundscape. *Area*, 26(3): 232–240.

Smith, S.J. (1997) Beyond geography's visible worlds: A cultural politics of music. *Progress in Human Geography*, 21(4): 502–529.

Smith, S.J. (2000) Performing the (sound)world. *Environment and Planning D: Society and Space*, 18(5): 615–637.

Taruskin, R. (1996) Punch into Pierrot (Petrushka). In: R. Taruskin (ed) *Stravinsky and the Russian traditions: A biography of the works through Mavra, vol 1*. University of California Press: Berkeley, Los Angeles, 661–778.

Tazudeen, R. (2020) The eco-sonic grotesque in Béla Bartók's The Wooden Prince. *Parallax*, 26(2): 179–194.

The Jim Henson Company. (2002) *The very best of The Muppet Show* (DVD). Columbia TriStar Home Entertainment: London.

Valentine, G. (1995) Creating transgressive space: The music of kd lang. *Transactions of the Institute of British Geographers*, 20(4): 474–485.

Walser, R. (2003) Popular music analysis: 10 apothegms and four instances. In: A.F. Moore (ed) *Analyzing popular music*. Cambridge University Press: Cambridge, 16–38.

Wood, N. (2012) Playing with 'Scottishness': musical performance, non-representational thinking and the 'doings' of national identity. *Cultural Geographies*, 19(2): 195–215.

Wood, N., Duffy, M. & Smith, S.J. (2007) The art of doing (geographies of) music. *Environment and Planning D: Society and Space*, 25(5): 867–889.

### *Discography*

Bartók, B. (1914–1916) *The Wooden Prince*. Neeme Järvi, The Philharmonia (recorded 1990). Chandos Records Ltd.

Birtwistle, H. (1967) *Punch and Judy*. David Atherton, London Sinfonietta (recorded 1979). NMC Recordings Ltd.

Cui, C. (1886) *Miniatures Op 39 no 1: Marionettes Espagnoles*. Maria Ivanova. Hänssler Classic.

Debussy, C. (1904) *Fetes Galantes Livre 1 no 2: Fantoches*. Irma Kolassi. Decca.

Goossens, E. (1917) *Kaledioscope, 18, 6: Punch and Judy Show*. Melbourne Symphony Orchestra. BC Classics.

Gounod, C. (1872/1879) *Funeral March of a Marionette*. Orchestra of the Royal Opera House. Decca.

Keith Harris and Orville. (1982) *I Wish I Could Fly*. BBC Records.

Macdowell, E. (1888/1901) *Marionettes Op 38*. Rudolph Ganz and Jeanne Behrend. Altair.

Martinů, B. (1914–1924) *Loutky (Puppets), I, II, III*. Paul Kaspar. Tudor.

Metallica. (1986) *Master of Puppets*. Electra Records.

Shaw, S. (1967) *Puppet on a String*. Pye.

Spitting Image. (1986) *The Chicken Song*. Virgin.

Stravinsky, I. (1911) *Petrushka* (1947 version). Klaus Tennstedt, London Philharmonic Orchestra (recorded 1992). London Philharmonic Orchestra.

Stravinsky, I. (1920) *Pulcinella (Suite)*. Otto Klemperer, Philharmonia Orchestra (recorded 1963, digitally remastered 1992). EMI Classics.

TryHardNinja. (2015) *The Puppet Song*. Songwriter: Gordiyenko Igor. FNAF SONG "The Puppet Song" (Animated Minecraft Music Video) – Bing video Accessed 26 March 2021.

Tyler, the Creator. (2019) *Puppet*. Columbia Records.

# 5   Theatrical puppets

In this chapter, my interest turns to possibilities arising from the use of puppets in specific theatrical settings. Given the long history of puppet theatre and diverse approaches to and spaces of puppet theatre, this chapter is in no way intended to provide an exhaustive review of theatrical puppets or puppets in theatre. Instead, I explore the role and value of puppetness for specific theatrical purposes, which is explored firstly through a targeted comparative analysis of two types of puppets to highlight their respective performative potential, and secondly through the lens of staged puppet performances that use puppets in place of animals to consider their affective potential.

My interest in this chapter, though, is not on the performances or on performance as a concept, but on performativity. Performance and performativity are understood in various ways, especially with respect to whether it is the performance or the performative that is characterised as re-inscription through repetition or a source of change and process of becoming (Shepherd 2016). While there is a whole body of work on performance theory, I do not digress into that literature in detail here, but instead adopt an established perspective on performativity within geography, in which performativity is generative of new possibilities, subjects and worlds, rather than reflective of the world (McCormack 2009) and need not be directly related to performance if performance is understood – for example – as a communicative act in public, a way of thinking about human behaviours or a practical form of thinking (Shepherd 2016). The minutiae of a practice not typically considered a performance can be performative in bringing about new possibilities for the further becoming of that unfolding practice and the entities involved in it, while a (for example) theatrical performance might not exhibit performativity. Thus, while I do consider specific performances with puppets, my interest is not in the performance as a performance but in which aspects of the puppet construction function performatively in generating new possibilities for puppet subjectivities and spatialities.

For the first analysis, I engage in a comparison of two stylistically similar but performatively very different types of puppets to highlight the specificity and potential of particular approaches to theatrical puppet performance. By highlighting the differences between the puppetry employed in The Muppets

DOI: 10.4324/9781003214861-11

(where the puppeteers are hidden from view) and that employed in productions such as *Avenue Q* (where the puppeteers are in plain sight), the specific ways in which these generate the body grotesque and the character contorted are drawn out, paying particular attention to the significance of faciality and pedality in establishing the believability of the scaling of the puppets, among other things. Yet despite these different bodily spaces, the consistency of the puppets' capacity to stimulate identification and empathy on the part of the audience is also underlined.

For the second analysis, I take this empathising potential as a springboard to explore the potential of puppet theatre to go beyond simple entertaining storytelling because of the affective power of puppets, especially when they exploit their own puppetness. Focusing on performances in which puppets are used as proxies for animals, the potential within puppets for narrative, dramatic and affective purposes is explored through a comparison of *Dr Dolittle* and *War Horse*, each of which is considered in relation to written text, film production and theatre show. This leads to further understandings of diverse approaches to the border and borderscape and to the identification of additional topological gulfs. Finally, this work is brought into conversation with ideas of the uncanny within puppetry literature – fluctuation between believing in the puppet and realising its artifice (Jentsch 1997; Cappelletto 2011; Gross 2011) – to suggest that the topological gulf between the border and the borderscape might be fruitful for understanding this peculiarly puppet phenomenon, enabling us to integrate otherwise slightly divergent perspectives.

Through these paired and highly targeted analyses, which focus on theatrical puppets but in the context of other cultural outputs, further diversity in configurations of the body grotesque and the character contorted is identified (analysis one), our appreciation of the various borderscaping practices through which these intercorporeal configurations are generated is advanced (analysis two) and the implications of this for our understanding of a peculiarly puppet phenomenon are outlined.

## Muppets versus puppets: performative bodily spaces

*Avenue Q*, with a mixed cast of human and puppet characters, was first performed in 2003 and it provides a valuable comparator to The Muppets due to both similarities and differences between the two types of puppets, even though *Avenue Q* is a staged production and The Muppets are associated with film and television. In terms of the visual style of the puppets, they both involve felt and fur-based bodily surfaces, brightly coloured complexions and ping-pong ball eyes. There are similarities, too, in the shows associated with each set of characters, from the name of the show (*Avenue Q*; *Sesame Street*) through the names of some of the characters (Trekkie Monster; Cookie Monster), to the more-or-less-ambiguous relationship between two of the characters (Rod and Nicky; Bert and Ernie). They also both contain strong moral messages, although as this is variable in the Muppet opus, given its longevity and diversity, my attention here is directed towards *The Muppet Christmas Carol* as an informative case study. While in *The Muppet Christmas Carol*, themes of charity, generosity and thankfulness are told through animalised social networks – for example, though the depiction of impoverished, hungry puppet mice and shivering, homeless puppet rabbits – in *Avenue Q*, themes of equality in relation to race and sexuality are told through a human–puppet social network that reveals prejudice to be equally problematic in both puppet and human communities. However, despite these similarities and the deliberate styling of the set design of *Avenue Q* in the vein of children's television, reminiscent of *Sesame Street*, there is no connection between the Henson Corporation (behind The Muppets) and *Avenue Q* (Anon 2007). Moreover, certain differences distinguish the two productions, in terms of both audience and puppet style, and these differences will bring us to the analytical focus for this first part of the discussion.

One difference relates to their respective audiences, which is child-focused for The Muppets, albeit with elements of humour that are appropriate to wider family viewing, compared to the explicitly adult content of *Avenue Q*, with the porn-obsessed Trekkie Monster and a sex scene involving a character named Lucy the Slut. The most important difference for our purposes, though, is the way in which they each use their puppets. One important point of divergence is the human–puppet relationship within the narrative. The human and puppet members of the population in *The Muppet Christmas Carol* are not distinguished from one another in narrative terms, being identified primarily as the characters that they play in Charles Dickens' story irrespective of whether the actor is human or puppet. While there is some acknowledgement that the puppets are acting as characters, including as Dickens himself, as in the joking between Gonzo (no discernible species) and Rizzo (a rat) as co-narrators who can foresee both forthcoming events and the climax of the story, there is still no identification of them as puppets. Rather, their depiction as actors serves to bolster their association with the human members of the cast, effectively denying their puppet status. By contrast, while the human and puppet characters interact in *Avenue Q*, there is much more explicit acknowledgement of differences between these two sets of cast members, with the question being asked at one point: 'Ready, normal people?' in a clear process of differentiating the puppet/human communities.

Another core difference between the two productions is the puppet–puppeteer relationship, which draws out further the starkly different ways in which puppets and humans are rendered equivalent in each production. Reflecting the contradictory ways in which puppeteers can engage with the illusion of puppetry – mask it or exploit it (Williams 1991) – The Muppets do the former by keeping the puppeteers and control mechanisms hidden from the viewer, but *Avenue Q* does the latter by revealing the control rods and making it plain that the facial expression of the human operator is part and parcel of the puppet character. Whereas The Muppets simultaneously remove the human operators from view and construct the puppets as human equivalents by directing attention to their characterfulness rather than their puppetness, *Avenue Q* simultaneously directs attention to the human operators through their exaggerated facial expressions and constructs the puppets as equivalents by becoming the facial expression for the puppet that they are operating. In other words, in The Muppets, the puppets performatively become human through the invisibility of the human, but in *Avenue Q*, the puppets performatively become human (and – reciprocally – the humans performatively become puppet) through the visibility of the human. Performatively, then, The Muppets achieve a unidirectional ontological shift through the evacuation of the human and denial of artifice, but *Avenue Q* achieves a bidirectional ontological merging through artifice.

This brings us to important insights as to what makes the puppets in *Avenue Q* so effective despite the obviousness of their artificiality, which I propose is related to interconnected issues of scale, faciality and (bi)pedality. In *The Muppet Christmas Carol* – although this film is far from unique in the Muppet library in this regard – puppet characters of all shapes and sizes interact with human-scaled characters and the Muppetised world that is depicted makes this plausible, so the viewer does not bat an eyelid when watching a spider do a deal on Scrooge's old bedsheets. The faces of the puppet characters are entirely Muppet in form and style and as their artificiality is masked by the removal of the human from the human–puppet assemblage, they are shown with whatever number of legs is appropriate to their form. As this pedality provides groundedness, the puppet character is scaled in relation to the ground, giving rise to small rat and rabbit characters interacting entirely unselfconsciously with much larger human characters. By contrast, although the *Avenue Q* set is scaled to 80% of human size, thereby unsettling normal scalar referents (Anon 2007), the puppets do not possess a body below the waist. The size of the puppets' heads is not that dissimilar to the heads of the human actors and their lack of legs means that the puppet characters can be operated at a similar height to that of the human actors. This – in turn – draws an equivalence in size and position even if not in complexion or appearance and incorporates the puppets within the community of human actors, scaled in relation to human head height. This is especially important to the effectiveness of these puppets precisely because of the incorporation of the facial expression of the puppeteer into the believability of the puppet character, as it is easier to merge the two sets of facial features in perceptual terms if the two (or more) faces are in close proximity. Counterintuitively, though, this lack of puppet legs does not result in a lack of groundedness for these puppet characters as this groundedness is secured through the interpenetration of human and puppet bodily presence and facial expression. While the puppet does not have its own legs, those of the puppeteer(s) fulfil the same function.

Seemingly, whether a puppet is operated by a single puppeteer or two, the perceptual merging of the facial expressions of human and puppet render the puppet–puppeteer entity a single whole that – perceived as a gestalt in this way – is deemed to have two legs. In a situation where there are two operators, the trunks of the puppeteers' bodies seemingly perform the function of the legs of a single operator. While a viewer might be capable of subitising the number of human legs and generate the rational and accurate number of legs to be four, the puppet character as perceived in gestalt terms has one face and two legs rather than three faces and four legs. The leglessness of the *Avenue Q* puppets, then, is a constitutive absence that is – from my perspective – fundamental to the effectiveness of these puppets in that they are both released from their own material groundedness to acquire human scalar relations and – through the facial convergence that this affords – perceptually bound into the corporeal materiality of the human puppeteers, thereby re-establishing their groundedness through the legs of the human operators. Performing at human head height, the puppets are enrolled into the company of human actors and the twin facets of facial proximity and expressive convergence collectively make them seem more real, while this enhanced expressiveness and bodily convergence with their human puppeteers reinforces this realism and imbues them with a stronger sense of innate vitality because they become perceived as one with their human operator/s rather than props to be manipulated. In this sense, then, the *Avenue Q* puppets are humanised. However, the attention directed towards the puppets by the human operators – through gaze, gesture and expressive congruence – foregrounds the puppet over the human performer/s, with the human vitality of the latter firmly invested in the former, effectively puppetising the human/s.

This constitutive absence, then, is crucial to the successful intercorporation of puppet and human bodies, which – in turn – is critical to establishing the vitality and believability of the puppet character despite the different elements (puppet, human) and the borders between them remaining in full view. Consequently, while faciality – and especially expressivity – is fundamental to the effectiveness of the puppets in *Avenue Q*, this does not operate in isolation but is bound up with a lack of puppet pedality and the scaling effects and alternative forms of groundedness that leglessness enables. Puppet bodies clearly do not need to be whole to be convincing, making apparent the artificiality of a puppet character does not need to objectify the puppet, and an effective puppet performance can be constituted as much by what it lacks (legs) as by what it possesses (a face).

These two approaches to puppet performance construct very different configurations of body grotesque and character contorted, as each either establishes or eliminates different topological gulfs in their varied practices of borderscaping. With The Muppets, even though puppets are archetypal examples of the grotesque, the body grotesque is denied (however bizarre the bodily form might be) because the human is performatively removed from the human–puppet assemblage, so the audience is only perceptually aware of the puppet as the character that it is playing. There is, though, contortion of character in The Muppets, but this is not malevolent, as it was in the narrative metaphorical constructions of puppets in Chapter 2. Instead, this takes the form of the Muppetisation of both the author of and characters within the original story. Charles Dickens becomes the blue hook-nosed Gonzo, and in the Muppetised Dickensian world, we see the introduction of a second Marley brother, enabling the brothers to be played by Statler and Waldorf (the two cantankerous old characters from the balcony in *The Muppet Show*) and the shift from Fezziwig to Fozziwig in honour of the Muppet bear Fozzi. However, herein lies an important distinction, as it is not the puppetisation of the original tale that introduces these adaptations, but the Muppetisation of the original tale. The telling of the tale is sculpted to fit the cast of the Muppet world, finding a significant role for key characters, even if this means inventing new characters or changing the names of existing characters. Counterintuitively, then, the Muppet-specific manner in which this tale is told need not be associated with puppets per se, even though Muppets are puppets, because their puppetness is not foregrounded and it is their individual characters rather than their puppet nature that is most important. This potentially reveals a topological gulf between Muppet and puppet that is specific to The Muppets, in that they are Muppets before they are puppets, but as they deny their own puppetness, the tension is not sustained and the gulf evaporates. Borderscaping in The Muppets is thus a three-way interaction between human, puppet and Muppet, which is oriented to groundedness to facilitate the believability of the inter-species and puppet–human relations that are portrayed and constructed at the level of the character rather than the body as the constitutive role of the human is denied, puppetness is not employed and they are Muppets first and foremost.

By contrast, the character contorted is largely absent from *Avenue Q*, although parallels can be drawn between the characters of *Sesame Street* and *Avenue Q*, reinforcing intertextuality as an additional consideration in this conceptualisation of the grotesque. However, the body grotesque in *Avenue Q* takes a unique form. Specifically, this involves an interesting conflict between a constitutive excess of puppet and human faces that become singularised into that of the puppet and a constitutive absence of puppet legs that become virtualised through those of the human operator/s in a distinct form of intercorporation. While faciality is clearly important here, this takes a different form compared to the emphasis on the ventriloquist's dummy in Chapter 1, being constructed on multiplicity rather than appearance, and through this facial borderscaping, the human performatively becomes a puppet. Simultaneously, this excessiveness of faciality is supplemented and contradicted by leglessness and through this constitutive absence of puppet legs, the puppet becomes human. Borderscaping in *Avenue Q*, then, is oriented to headedness to facilitate the characterisation of the puppets as well as the believability of human–puppet relations, and is constructed at the level of the body, establishing a bidirectional intercorporation at the levels of head and legs across two particular topological gulfs between head and legs and between excess and absence, which is supplemented by a third between singularity and multiplicity.

Emerging from this comparative analysis are two distinct sets of borderscaping practices, one more attentive to character and the other more attentive to the body, as although the body grotesque is precluded in The Muppets it is foundational to *Avenue Q*. Each sets up and resolves its own topological gulfs to enact its specific form/s of intercorporation, which – in turn – generates very different human–puppet spatialities. Yet each set of borderscaping practices constructs puppets that are both effective and affectively powerful, to which it is easy to relate and with which it is very easy to empathise, suggesting that puppet performances can function as effective vehicles for moral messaging due to their affective power. It is precisely this potential that forms the analytical focus for the second part of this chapter, through a focus on puppet associations with animals.

## Animal puppets, moral messaging and affective power

As outlined in Chapter 1, the puppet productions receiving the highest acclaim are those deemed to deliver the most expressive, realistic and graceful movement (Billington 2009, 2015; Hitchings 2018; Herring 2019), but notably, and as will be explored, the affective power of these performances is intense despite the deliberate exposure of the craft and artistry behind the puppet characters (Hitchings 2018), stimulating warnings that audiences for *Running Wild* will end up in tears despite knowing that it is just a puppet on a pole (Norman 2016). Also of significance for this chapter is a sense of the prioritisation of the animal puppet characters over the human dimension of the story being told, introducing a broader more-than-human perspective beyond the human–puppet relation. This discussion delivers a targeted comparative analysis of two puppet stage productions that both use puppets as proxies for animal actors – *War Horse* and *Dr Dolittle* – and that both convey strong moral messages: Animal rights (*War Horse*) and environmental protection (*Dr Dolittle*). The first of these puppet productions is a popular and long-running show and a box office hit but the second was closed months early after only its third tour venue due to low advance ticket sales and has been described as a disappointing failure (Maxwell 2019). I have selected these two productions for examination here partly because of their different fates despite similarity in their uses and styles of puppetry, and partly because of their different engagements with puppetness despite these similar uses and styles of puppetry.

I analysed three versions of each work:

1. *Dr Dolittle*: The original story by Hugh Lofting, the 1967 film version with Rex Harrison and a 2019 theatre production involving puppets.
2. *War Horse*: The original story by Michael Morpurgo, the 2012 Steven Spielberg film version and the 2017 theatre production involving puppets.

Both *War Horse* and *Dr Dolittle* are explicitly attentive to human–animal relations and are robust in their moral messaging, yet this messaging is also undermined by factors both independent of and related to their use of puppets. Both productions draw out a contrast between convivial human–animal relations of companionship and the objectification, commodification and brutalisation of animals, but plot considerations undermine the potential power of this contrast, aside from the limitations imposed by the use of puppets as proxies for animals. The stage production of *War Horse*, for example, presents the reunion between Albert and his horse, Joey, as incidental rather than a pursued goal, thereby trivialising the human-horse relationship and undermining the narrative foundation in which the moral message is grounded. In the stage production of *Dr Dolittle*, which was explicit in its advocacy of vegetarianism and involved a request for funding for conservation purposes in both the programme (Anon 2018) and at the end of the show, the Pushmi-Pullyu continued to be seemingly unproblematically commodified as a source of income and the fantastical creatures serve as a form of transportation for the human characters, so there is an inconsistency between the message of the tale and the tale itself.

Moreover, while the proxy use of puppets in place of animals avoids accusations of animal exploitation (the 1967 film version of *Dr Dolittle* included costumed animals performing in a circus and a dressed chimpanzee 'cooking' dinner over a stove), this use of puppets precludes more extensive engagement with the performativity of puppets, which could bolster the strength of the moral message. In these works, animals are puppetised for utilitarian rather than narrative purposes, and relations between puppets and animals – or puppets and humans – are left largely unexamined in favour of the relation between humans and animals. In both *War Horse* and *Dr Dolittle*, animals are puppetised and a core theme in each story is human–animal relationality, but especially in *Dr Dolittle* the fact that the story is being told through characters that are puppets rather than animals is treated as a given rather than a source of inspiration for critical social commentary, so there is potential to make greater use of the materiality and artifice of the puppets in conveying the moral message at the heart of a story as well as the story itself, which might provide a means of avoiding reviewer criticism for overly earnest or heavy-handed moral messaging through the tale itself (see, for example, Gardner 2016; Mountford 2016; Norman 2016).

Notably, these criticisms were directed towards *Running Wild*, a production highly esteemed for puppetry skill (Gardner 2016), whereas *The Lorax* was acclaimed for both its artful puppetry and the subtlety of its moral environmentally friendly message (Billington 2015). Seemingly, the perceived effectiveness of the moral messaging is not tied directly to either the puppetry skills involved in telling the story or the realism of the puppet characters, suggesting that interrogating puppetness might be a productive avenue for embedding moral messages into puppet productions without being overly blatant in its narrativisation. Whereas in literary metaphorical uses, the puppetisation of the human entailed the deindividualisation of the victim, in these performative uses, the puppet character is not only animated but individualised as they are granted unique personalities and motivations, accentuating the need for and target of empathy. This, though, risks undermining the potential to exploit their puppetness if the puppets are treated as proxies for animals rather than as material-subjective resources for dramatic exploration, leaving the narrative to carry the load of the moral message. There is, then, the potential for the use of puppetness (as a quality) – distinct from puppets (as entities) – in conveying moral messages and an additional way in which we can think of the utilitarian dimension of the dimensional framework – social or moral messaging – which can but might not engage puppetness explicitly in doing so.

However, *War Horse* does make very effective use of puppetness for dramatic and affective purposes, and a close examination of this use of puppetness will enable further development of the dimensional framework and of the borderscape by alerting us to diverse approaches to its treatment, thereby establishing the gap between the border and the borderscape as its own topological gulf. Further, this discussion will be brought into conversation with existing work on puppets as uncanny (Jentsch 1997; Cappelletto 2011; Gross 2011), to integrate certain differences within existing work in this area, but the starting point is an exploration of the dramatic and affective use of puppetness, which is explored here in relation to: Scale and atmosphere, more-than-human capacities, and deconstruction and artifice.

A range of puppet types are used in the stage production of *War Horse*, including birds on sticks, a goose that trundles around on a wheel and horses that are operated by multiple puppeteers. With all these puppets, the puppeteer is clearly visible, although those operating a horse's legs from within the horse's body swiftly blend into the overall perception of the horse. That said, the excessive number of legs compared to the number of animals being represented helps to generate a sense of scale, crowdedness or confusion to establish an atmosphere appropriate to the scene being performed (e.g., a battlefield) that would otherwise be elusive in the relatively small environs of a theatre stage. These puppet practices, then, are consistent with the theatrical dissolution of the human/puppet border proposed in Chapter 3, as the border is partially concealed from view while still being visible, thereby generating dramatic effects.

It is the horse puppets that have received the greatest attention as they are especially impressive in their scale, expressiveness and movement, and they are used in different ways for powerful affective impact. One of the key affordances of puppets is their material capacity to withstand brutality, and this more-than-human capacity makes puppets ideally suited to narratives and performances that involve maltreatment of the characters they represent. However, the power of a convincing puppet means that the affective impact on the audience is not reduced by the puppet nature of the thing being brutalised. In buying into the reality, sentience, individuality and experience of the puppet character, the impact of that treatment in affective terms is conveyed powerfully, and even more so as the knife that is driven through the head of the puppet horse to end its misery does not have to be fake. A real knife can be driven through a real (puppet) head, yet the puppet is so convincing that it is perceived as being driven through a real (horse's) head. This is a radical dissolution of the border between horse and puppet, rather than between human and puppet, as the puppet horse is pretending to be a real horse, but the audience perceives it as a real horse and ignores its status as pretence. The reality of the action of driving a knife through the head reinforces the sense of reality on the part of the character being brutalised (euthanised) even as the reality of the act is only enabled by the pretence of the puppet.

Another performative benefit of using puppets is their capacity for deconstruction. In *War Horse* this is evident in two different scenarios, both of which were highly affectively charged and both of which – counterintuitively – functioned by drawing attention to the artifice of puppetry rather than concealing it, thereby theatrically strengthening (van Houtum and Eker 2015) the human/puppet border. The first such instance involved the separation of a horse from its rider during a battle scene, in which both the horse and its rider were puppets and in which the control rod for the rider, extending from the bottom of the rider's torso downwards, became clearly evident in silhouette. Rather than reducing the affective power of this scene, however, the transparency of the objectivity of the rider made the scene more powerful. It both emphasised the reduction of human vitality to base materiality in death and threw into sharp relief the capacity for bodies to be torn apart in battle. The rider puppet did not need to be whole to be effective when on the horse and – once off the horse – the fact that it was not whole was foundational to the affective power of that scene. Consequently, although puppet performances are often judged on the basis of how convincing the puppets are in their movement, appearance and expression, there are also times when their artificiality is not only put to dramatic use but is also fundamental to its affective power.

The second instance at which the audience's attention was drawn to the artificiality of the puppets involved the death of Topthorn, Joey's primary horse companion. At this point in the production, the puppeteers who had been operating Topthorn from inside withdrew from the physical form of the puppet and stood in still silence, heads bowed in honour of the fallen horse hero. This explicit separation of puppeteers from puppet highlighted the artificiality of Topthorn and seemingly goes against the emphasis in much puppet work on making the puppeteer invisible or at least barely noticeable, but again, the affective power of this scene was magnified through this display of artifice rather than trivialised by it. The withdrawal of the puppeteers from Topthorn's puppet body was a physical departure of that which had given him his vitality. At this point, the emphasis on the puppetness of Topthorn emphasised the source of his vitality and the departure of this source of vitality further emphasised the deathly inertness of Topthorn. Even when physically separated from the puppet that they had energised, the puppeteers were crucial to the believability of the puppet as a (formerly) living being, as the affective power of the human–puppet intercorporation endured beyond the physical enactment of intercorporation. Despite the bodily intercorporation being broken, both a perceptual intercorporation and an affective intercorporation remained. Through both a theatrical strengthening and physical

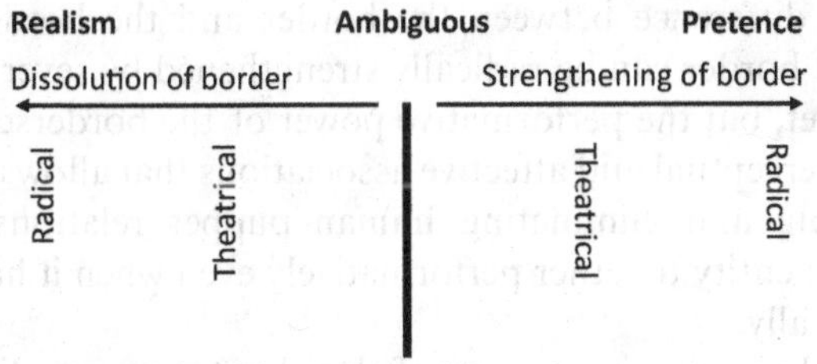

| Realism | Ambiguous | Pretence |
|---|---|---|
| Dissolution of border | | Strengthening of border |

*Figure 5.1* Human/puppet border treatments.
Source: Author.

extension of the human/puppet border, transparency as to the puppetness of Topthorn contributed to the affective power of his death by enhancing the former vitality of the puppet. This combination of both strengthening and extending the border could be considered a radical strengthening of the border, more powerful than either strengthening or extending the border individually. Here, again, then, is a constitutive absence, as the withdrawal of the puppeteers from the bodily form of the puppet horse starkly reveals the void – the topological gulf – between the puppet's form and its source of vitality, which magnifies the belief in its former vitality and the sadness at its loss. Consequently, puppetness is employed for dramatic purposes in *War Horse* in different ways – as illustrated in Figure 5.1 – with astonishing affective power.

## Borderscaping and the uncanny

Revealing the human/puppet border in *War Horse* serves to enhance the affective investment of the audience in the puppet despite revealing the artifice behind it. Drawing on and extending similar ideas related to political borderscapes (van Houtum and Eker 2015), while the border between puppet and human in the performative use of puppets can be theatrically dissolved, as in the generation of scale and confusion, that same border can also be radically dissolved, as in driving the knife through the horse's head. Furthermore, it can be theatrically strengthened, as with the rider's torso; and/or radically strengthened, as in the death of Topthorn. These additional perspectives supplement the previous conceptualisation of the theatrical dissolution (see Chapter 3) of the border by diversifying the ways in which human–puppet relations can be sustained, eliminated, ambiguated or emphasised for dramatic and affective purposes. Whether dissolved or strengthened, the distinction between theatrical and radical rests on the extent to which the visibility of the border changes, e.g., through strengthening it, extending it, or both, which can be affectively powerful.

This also establishes the borderscape as a crucial yet malleable space of puppetness which can itself be used for both narrative and performative purposes even in the erasure of the border at its heart. Moreover, it highlights the difference between the border and the borderscape as the human/puppet border can be radically strengthened by severing the human from the puppet, but the performative power of the borderscape persists as a network of perceptual and affective associations that allows for simultaneously sustaining and eliminating human–puppet relations, holding the human–puppet entity together performatively even when it has been deconstructed physically.

As with the discussion in part one of this chapter, we can discern yet more permutations of topological gulf, intercorporation and borderscaping in the construction of puppets. Most significantly, these three – topological gulf, intercorporation and borderscaping – come together through consideration of the topological gulf drawn out between materiality and vitality in the discussion of *War Horse*. While the two examples – the separation of rider and horse and the withdrawal of the puppeteers from Topthorn – are similar in playing with that topological gulf, they are also distinct in that the topological gulf related to the rider highlights the materiality of the human whereas that related to the horse highlights vitality, even spirituality, evidencing flexibility in how the same topological gulf can be employed and reinforcing the elevation of the subjectivity of animals above that of humans in these works.

Moreover, these examples can advance our understanding of the relationship between the border and the borderscape as its own topological gulf, revealed through the dismantling of the intercorporation that had – for example – animated Topthorn. The material deconstruction of Topthorn tore apart humans and puppets in a radical strengthening of the border that simultaneously separated the bodily form from the source of its vital spirit. However, the affective and subjective reality – the spirit – of Topthorn as a previously living being is sustained beyond its death by the borderscape that still holds its human and puppet elements in creative and unified tension despite their separation. The spirit – the source of vitality – has been liberated from the form of the body by the withdrawal of the human operators but is still palpable within the broader relational and affective borderscape because those separated human operators are still bound to the bodily form in their mourning, and its palpable endurance within the borderscape amplifies its absence from the form of the body.

It is at this point that puppets as uncanny become relevant. While there is no single scientific label for the phenomenon or experience of alternately believing and disbelieving in the life of a puppet, many names have been attributed to it (Francis 2012), with reference to the uncanny being a common feature. Oscillation, opalisation and double-vision (see, for example, Gross 2011; Jurkowski 2013) have all been utilised to refer to shifting perceptions of a puppet as inanimate and as living, as subject and as object, making us believe and not believe at the same time (Bicât 2007). Indeed, the ontological paradox of being both a material object and a signifier of life is the inescapable tension within the puppet itself (Tillis 1996). However, distinctions are discernible between authors who seem to emphasise an alternation between these perceptions (Zamir 2010; Gross 2011; Francis 2012) and those who describe this phenomenon more in terms of a simultaneous duality, a tension or a suspension midway between the alternatives (Tillis 1996; Bicât 2007; Jurkowski 2013). Bicât, for example, explicitly refers to believing and disbelieving at the same time, which seems highly consistent with the notion of double-vision, not as an alternating perception but a simultaneous conflicted perception. My intention here is not to overstate these differences, nor is it to inject yet more labels for this phenomenon, but rather, to work through how the notions of borderscaping and topological gulf can help to unify some of these different perspectives on it.

Picking up once again on the thread of the theatrical/radical strengthening/dissolution of the border between puppet and human, a dissolution of the border that integrates material object and subjective being more effectively or convincingly would lend itself to ideas of the puppet being held in suspension between the two ontological extremes, whereas a strengthening of the border that makes the discreteness of each element more apparent would suggest that alternating perspectives are more appropriate. With the latter, there is – of course – a chance that the border is now so strongly drawn that no alternation or oscillation is possible, and the phenomenon breaks down. However, this is where the borderscape becomes most important, as this broader set of perceptual and affective connections across the border can sustain residual investment in the puppet even though the border at its heart has been torn apart. Negotiating the tension between material object and subjective being in the puppet itself, then, is scaffolded by the tension between the border and the borderscape. On this reading, the most appropriate description of this phenomenon is seemingly the disintegration and regeneration of unity (Jurkowski 2013), wherein the unity of significance is that of the borderscape rather than that between the two elements being brought together or pulled apart at the border. A breakdown in the tension between material object and subjective being brought about by the radical strengthening of the border can leave the puppet disintegrated at the border, but the maintenance of unity within the borderscape can facilitate the resurrection of oscillation and the reconstruction of unity across the border. Providing that the borderscape itself does not break down entirely, the potential remains for the puppet to be re-enlivened.

Thus, not only can the revelation of the artifice lead to enhanced rather than diminished investment in the reality and subjectivity of the puppet, but the suspension of disbelief can be sustained and reinstated despite that revelation of artifice through the scaffolding of the borderscape. This topological gulf between border and borderscape, which is in both tension and flux in puppet performance, is perhaps significant for understanding – and for integrating different understandings of – puppets as uncanny. This is especially so in the context of Jurkowski's (2013) articulation of this as the disintegration and regeneration of unity, as it perhaps allows us to understand the enchanting power of the puppet even as its artifice is exposed, by providing a mechanism for how the liveliness of the puppet incorporates something of its lack of life (Gross 2011).

## Conclusion

This chapter has expanded consideration of the varied uses of puppets to explore their use in specific theatrical contexts and for specific purposes. Through a comparative consideration of two different styles of puppet performance – The Muppets and *Avenue Q* – two divergent approaches to borderscaping were identified, each of which constitutes and resolves (or not) its own topological gulfs in diverse forms of intercorporation that generate particular bodies grotesque, characters contorted and spaces of puppets. While the body grotesque was found to be denied in *The Muppet Christmas Carol* by the performative removal of the human from the puppet–human assemblage, intercorporation functioned at the level of the character with the character of the human being puppetised (or more accurately, Muppetised), which highlighted a topological gulf between Muppet and puppet that is constitutive of that character contorted. The intercorporation within *Avenue Q* functioned at the level of the body and operated both at two levels and in two directions. Through the perceptual blending of multiple faces into a singularised faciality of the puppet and the perceptual filling of the constitutive absence of puppet legs, humans become puppetised and puppets become humanised. In this way, leglessness is a constitutive absence that supplements the constitutive excessiveness of faciality and both leglessness/faciality and excessiveness/absence constitute topological gulfs through which these bodies grotesque are created. In terms of the spatial generativity of these works, The Muppets' world is oriented around the puppets (in unique Muppet form) and this world is scaled in relation to groundedness, but as they are Muppets before they are puppets, puppetness itself contributes little to this spatiality. By contrast, the world of *Avenue Q* is oriented around humans and this world is both downscaled (to 80% of human proportions) and scaled in relation to headedness. The intercorporation of humans and puppets through bidirectionality at the level of both heads and legs means that puppetness is fundamental to the generation of both the spatialities and subjectivities that characterise this production.

This chapter has also introduced a more-than-human perspective by considering the incorporation of animality as well as humanity into puppet performances. Three versions each of *War Horse* and *Dr Dolittle* were analysed, with puppets and puppetness being used in different ways and to different degrees between the works, from puppets performing as animal proxies in *Dr Dolittle* to puppetness generating affective impact in *War Horse*. While the use of puppets as proxies limits their potential for engaging fully with puppetness in narrative terms, it was suggested that puppetness could be employed to enhance the impact of a tale's moral message. *War Horse*, though, did make very effective use of puppetness for dramatic and affective purposes, drawing out a topological gulf between vitality and materiality, whereby the horse's spirit palpably endured within the borderscape despite – and, indeed, because of – the revelation of the artifice through the separation of puppet and puppeteer at the border.

In terms of the spatial generativity in these works, the use of puppets for dramatic effect in *War Horse* revealed not only a further utilitarian possibility in the exploitation of puppetness for affective impact but also the multiplicity of the borderscape as containing a border between puppet and puppet (as in the case of the separation of the rider from the horse) as well as between puppet and puppeteer (as in the death of Topthorn). This, in turn, highlighted the affective power that can be generated by accentuating rather than concealing the artifice behind the puppets, which – conversely – reinforced the affective believability of the puppet character. By considering various ways in which the human/puppet border can be radically dissolved, theatrically dissolved, theatrically strengthened or radically strengthened, we can appreciate how the human/puppet border is simultaneously sustained and eliminated, through contributing to the affective power of the borderscape even while its physical integrity is radically ruptured at the border. Moreover, as its own topological gulf, this relationship between the border and borderscape can help sustain and reinstate the believability of the puppet despite the radical strengthening of the border, enabling us to accommodate and integrate multiple and diverse perspectives on puppets as uncanny within a broader notion of the disintegration and regeneration of unity within the borderscape.

Ultimately, this chapter develops our understandings of puppets and how they are variously constructed by drawing out new dimensions from this comparative consideration of different tellings of the same cultural tale, which provide more detail on the oxymoronic process of simultaneously sustaining and eliminating the conjunction between human and puppet for narrative, moral, performative and ontological purposes. It also furthers our appreciation of the diverse bodies grotesque and characters contorted that can be constructed through different approaches to the border/s upon which they are grounded, deepens our engagement with the topological gulf and the borderscape and speaks to concerns in the existing literature on puppetry to integrate different perspectives on puppets as uncanny. As the final chapter in Part 2, the analytical emphasis has also started to progress to ideas of puppets becoming human and humans becoming puppet (e.g., the bidirectionality identified through faciality and pedality in *Avenue Q*), which signals the forthcoming emphasis on constructions of puppet–human transformations in Part 3.

## References

Anon. (2007) *Avenue Q*, programme.

Anon. (2018) The value of nature. *Doctor Dolittle: The musical, souvenir brochure*. Encore International Merchandising, n.p.

Bicât, T. (2007) *Puppets and performing objects: A practical guide*. The Crowood Press: Marlborough.

Billington, M. (2009) War Horse. *The Guardian*, 06 April 2009. https://www.theguardian.com/stage/2009/apr/05/theatre-review-war-horse. Accessed 18 October 2020.

Billington, M. (2015) Dr Seuss's The Lorax review – the best family show since Matilda. *The Guardian*, 17 December 2015. https://www.theguardian.com/stage/2015/dec/17/dr-seuss-the-lorax-old-vic-london-family-show-david-greig. Accessed 18 October 2020.

Cappelletto, C. (2011) The puppet's paradox: An organic prosthesis. *RES: Anthropology and Aesthetics*, 59/60: 325–336.

Francis, P. (2012) *Puppetry: A reader in theatre practice*. Palgrave Macmillan: Basingstoke.

Gardner, L. (2016) Running Wild Review: Michael Morpurgo animal magic rivals War Horse. *The Guardian*, 22 May 2016. https://www.theguardian.com/stage/2016/may/22/running-wild-review-michael-morpurgo-puppets-war-horse. Accessed 18 October 2020.

Gross, K. (2011) *Puppet: An essay on uncanny life*. University of Chicago Press: Chicago, IL.

Herring, N. (2019) Getting in the saddle for an emotional gallop. *Oxford Times*, 29 August 2019, p34.

Hitchings, H. (2018) War Horse Review: Heart-tugging, visually inventive spectacle is back in the saddle. *Evening Standard*, 09 November 2018. https://www.standard.co.uk/go/london/theatre/war-horse-national-theatre-review-a3985721.html. Accessed 18 October 2020.

van Houtum, H. & Eker, M. (2015) Borderscapes: Redesigning the borderline. *Territorio*, 72: 101–108.

Jentsch, E. (1997) On the psychology of the uncanny. *Angelaki*, 2(1): 7–16.

Jurkowski, H. (2013) *Aspects of puppet theatre*, 2nd edition. Palgrave Macmillan: Basingstoke.

Maxwell, D. (2019) Review: Dr Dolittle at the New Theatre, Oxford. *The Times*, 18 January 2019. https://www.thetimes.co.uk/article/review-dr-dolittle-at-the-new-theatre-oxford-jkw8cg38r. Accessed 20 October 2020.

McCormack, D.P. (2009) Performativity. In: R. Kitchin & N. Thrift (eds) *International encyclopedia of human geography*. Elsevier (online resource), 133–136. Accessed 11 July 2021.

Mountford, F. (2016) Running Wild, theatre review: Creature comforts. *Evening Standard*, 20 May 2016. https://www.standard.co.uk/go/london/theatre/running-wild-theatre-reviews-creature-comforts-a3253001.html. Accessed 20 October 2020.

Norman, N. (2016) Theatre reviews: Monster Raving Loony and Running Wild. *The Daily Express*, 26 May 2016. https://www.express.co.uk/entertainment/theatre/674165/Theatre-reviews-Monster-Raving-Loony-Running-Wild. Accessed 20 October 2020.

Shepherd, S. (2016) *The Cambridge introduction to performance theory*. Cambridge University Press: Cambridge.

Tillis, S. (1996) The actor occluded: Puppet theatre and acting theory. *Theatre Topics*, 6(2): 109–119.

Williams, M. (1991) Aspects of puppet theatre/the language of the puppet. *Australasian Drama Studies*, October 1, 19: 67–75.

Zamir, T. (2010) Puppets. *Critical Inquiry*, 36(3): 386–409.

**Books, films and shows analysed**

*Avenue Q*, New Theatre, Oxford (2014).
*Dr Dolittle* (1967) Twentieth Century Fox. Dir: R. Fleischer.
*Dr Dolittle*, New Theatre, Oxford (2019).
Lofting, H. (1920/2013) *The Story of Doctor Dolittle*. Vintage Classics: London.
*The Muppet Christmas Carol* – 50th Anniversary edition (2005), Disney/Jim Henson Productions. Dir: B. Henson.
Morpurgo, M. (2014) *War Horse*. Egmont: London.
*War Horse* (2012) Touchstone Home Entertainment. Dir: S. Spielberg.
*War Horse*, New Theatre, Oxford (2017).

# Conclusion to Part 2

This second and central part has developed the idea of borderscaping as the means by which unique and peculiar puppet geographies are brought into being through targeted interrogation and exploitation of the human–puppet bodily border, and through that discussion, diverse approaches to, and forms of, the body grotesque and character contorted have been drawn out. Chapter 3 analysed diverse approaches to the use of puppets in literature and film to introduce a dimensional framework and forge an initial conceptualisation of the borderscape. Chapter 4 attended to musical constructions of puppets and highlighted both mood and movement and contrasts and tensions as key considerations. Chapter 5 focused on theatrical uses of puppets for moral and affective purposes to extend and refine both the dimensional framework and the practice of borderscaping, which was then related to understandings of puppets as uncanny. In this conclusion, I briefly recap on the findings of each chapter through a comparative discussion, before bringing the outcomes of these chapters together to establish a consolidated outcome. Specifically, this consolidation includes refining the dimensional framework to encapsulate these developments, responding directly to the question of what grants puppets their performativity and considering how well current formulations of the relationship between the cognitive and the affective in non-representational geography can accommodate the uncanniness of the puppet.

## Comparative analysis

The analysis in Chapter 3 teased out six dimensions – utilitarian (role, extent), relational (individual, social), qualitative (puppetness, ontology) – found to characterise different approaches to the use of puppets and developed these into an analytical framework to support and stimulate subsequent research. A detailed investigation of space in these works identified a preponderance of certain everyday spaces, alongside a graduated reworking of everyday spaces into spaces of puppetness as the nature of a puppet and the human–puppet relationship are increasingly employed as focal factors in the telling of the story. Moreover, the significance of bodily spaces and spaces of memory was highlighted in the generation of these peculiarly puppet worlds, whereby the most interesting puppet worlds emerge from the integration of

DOI: 10.4324/9781003214861-12

puppet–human conjoined bodies, conflated spaces and conflicted memories. The bodily and psychological aspects were respectively associated with the body grotesque and the character contorted, which were found to take varied forms and to occur either independently or in combination. The notion of intercorporation was proposed to denote this integration, which was compared to and distinguished from performative ideas of transembodiment (Mello 2016). These puppetised geographies were resituated in the disciplinary context by developing and integrating concepts from comic book geographies: Borderscapes (dell'agnese and Amilhat Szary 2015; van Houtum and Eker 2015) were redirected to the bodily border between puppet and human, and the forging of relations across the topological gulf (Dittmer 2010; Dittmer and Latham 2014) was compared to the investment in the intercorporation. These were brought together in the specification of intercorporation as a theatrical dissolution of the puppet–puppeteer border through the simultaneous elimination and reinstatement of puppet–puppeteer distinctiveness.

Chapter 4 engaged with geographies of music, outlined varied ways in which puppets are associated with music and musicians and presented an analysis of lyrical and musical constructions of puppets from the perspective of a musically untrained listener. This revealed strong consistency with previous constructions of puppets, such as a predominance of the marionette as the type of puppet constructed, but also drew out differences, such as the lack of facial specificity. Two overarching themes emerged through the analysis: Mood and movement, and tensions and contrasts. The construction of puppets through evocations of mood and movement was articulated as twin perceptions of grounded lightness generated through exaggerated jauntiness ('plinky-plonkiness') and anatomical incoherence through either disassembly or solidification of an otherwise multi-limbed anatomy ('perky-jerkiness'). Tensions and contrasts included those between humour and sadness, gentility and abrasiveness, and festive and ominous, but a further important tension – which was related to mood and movement – emerged between expectation and experience, all of which were associated with the contrariness of puppets. These features were especially notable with respect to Birtwistle's construction of Mr Punch, which was deemed to have configured a combined body grotesque and character contorted through the evacuation of Mr Punch's subjectivity and distribution of his character to the percussive background rather than his personal lyrical and musical articulations, constructing Mr Punch as hyper-puppetised as an archetype but de-individualised as a character.

Chapter 5 explored the use of puppets in specific theatrical contexts, to dig more deeply into the significance of different types or styles of puppets for the construction of the body grotesque and the character contorted and to explore the potential and power of puppets for moral messaging and affective impact. A comparative analysis of The Muppets and *Avenue Q* found that the body grotesque was denied in The Muppets through the removal of the human from view and intercorporation at the level of the character, as they are constructed as Muppets rather than puppets. By contrast, the body grotesque in *Avenue Q* involves a bidirectional human–puppet convergence anchored on the face and the feet, involving an excess of faciality and a constitutive absence of legs. Subsequently, three versions each of *War Horse* and *Dr Dolittle* were analysed, with puppets performing as animal proxies in *Dr Dolittle* but puppetness generating affective impact in *War Horse*. Exposition of the potential to employ puppetness to communicate and emphasise the impact of a tale's moral message identified an additional aspect to the utilitarian dimension of the framework, while the use of puppetness for dramatic and affective purposes in *War Horse* revealed not only a further utilitarian possibility in the exploitation of puppetness for affective impact but also the multiplicity of the borderscape as containing a border between puppet and puppet (as in the case of the separation of the rider from the horse) as well as between puppet and puppeteer (as in the death of Topthorn). This, in turn, highlighted the affective power of accentuating rather than concealing the artifice behind the puppets, conversely reinforcing the affective believability of the puppet character. Moreover, the relationship between the border and the borderscape was established as a topological gulf and it was suggested that this might help us to understand or conceptualise the oscillation or double-vision – the conflict between believing and disbelieving its vitality – that characterises puppets as uncanny, as the affective power of the borderscape can be sustained even while its constitutive border is radically ruptured, which can be strategically and variously used for narrative, moral, dramatic, affective and ontological purposes.

Collectively, these chapters have proposed, tested, affirmed and further developed a dimensional framework of cultural uses of puppets and have established a reconfiguration of the borderscape to conceptualise a continuum of approaches to the oxymoronic treatment of the border that can sustain the affective power of the borderscape even while the border itself is radically strengthened. Consequently, the dimensional framework from Chapter 3 requires revision to incorporate the moral and performative utilitarian dimensions of puppets and the varied approaches to the borderscape identified in Chapter 5, and the distinction between the body grotesque and the character contorted that has been consolidated in these three chapters. In Figure C2.1, overleaf, the moral utilitarian dimension forms a continuum from puppets and puppetness not being used to bolster the moral message within the tale, through puppets but not puppetness being used to do so and/or doing so only indirectly or unintentionally, to puppetness being directly employed to strengthen the impact of the moral message within a story. The performative utilitarian dimension establishes a continuum from puppet effects in evidence but not being used for specific dramatic purposes, through their use for specific effects but without exposing the artifice behind them, to deliberate exposure of that artifice as a means of establishing dramatic effect and affective power. Similarly, borderscape has been included as an additional relational dimension from the maintenance of the bodily conjunction between puppet and puppeteer, through the deletion or denial of any bodily

| | | Level 1 | Level 2 | Level 3 |
|---|---|---|---|---|
| **Utilitarian** | Role / *Proxy* | **Incidental**: Puppets rarely feature | **Indirect**: Puppets and humans used similarly | **Focal**: Puppets used instead of humans |
| | *Narrative* | **Incidental**: Puppets as unquestioned objects | **Indirect**: Puppets provide new perspective but not intentional or central | **Focal**: Puppets questioned as a narrative device |
| | Impact / *Moral* | **Incidental**: Puppets not used for moral messaging | **Indirect**: Puppets used indirectly for moral messaging | **Focal**: Puppetness targeted for moral messaging |
| | *Performative* | **Incidental**: Puppets not used for dramatic impact | **Indirect**: Puppet effects used but artifice not exposed | **Focal**: Artifice exposed for dramatic/affective impact |
| | Extent | **Minimal**: Puppets not a common or major aspect of the work | **Partial**: Puppets feature significantly but not a/the driving force | **Full**: Puppets used extensively and thoroughly, the main feature |

*Increasing integration* →

| | | Level 1 | Level 2 | Level 3 |
|---|---|---|---|---|
| **Relational** | Individual | **Informational**: Puppet as simply part of the world factually | **Social**: Puppet as social mediator, modifying social acceptability | **Psychological**: Puppet as communicative facilitator |
| | Social | **Discrete**: Puppet and human societies distinct | **Metaphorical**: Puppet society as metaphor for human life | **Constitutive**: Puppets humanised, humans puppetised |
| | Borderscape | **Conjoined**: Physical connection made/maintained | **Deleted**: Physical connection erased/denied | **Separated**: Physical connection actively broken |

*Increasing interrogation* →

| | | Level 1 | Level 2 | Level 3 |
|---|---|---|---|---|
| **Qualitative** | Puppetness | **Negligible**: No consideration of puppet nature | **Moderate**: Some consideration of puppet nature | **Substantive**: Significant exploitation of puppet nature |
| | Ontology | **Conventional**: Narrative remains within human context | **Creative**: Modification of/addition to human context | **Transformational**: New worlds created merging puppets and humans |
| | Grotesquery | No grotesque elements | Either body grotesque or character contorted | Both body grotesque and character contorted |

*Increasing intercorporation* →

*Figure C2.1* Revised dimensional framework.

Source: Developed from Banfield 2020.

borderscape, to the deliberate severance of the connectivity of the border-scape for performative reasons. Finally, the potential for the body grotesque and the character contorted to be either neglected, constructed individually or configured in combined form has been incorporated within the qualitative dimension.

## Affect, the borderscape and the uncanny

Part 2 has been concerned with the question 'what grants puppets their per-formativity?', which has been explored not from the perspective of perfor-mance with puppets but from the perspective of how puppet constructions are received and perceived by the (this) viewer/listener. Emerging from these three chapters are key sources of this performativity: Explicit interrogation of puppetness, the borderscape as an active arena for this interrogation and different approaches to this practice of borderscaping (theatrical–radical dis-solution or strengthening of the border/s at its heart), which were found to generate powerful affective impacts on the part of the viewer. Indeed, affect is a recurrent or underlying theme throughout these three chapters, with Chapter 3 acknowledging poignancy, fear and repulsion, Chapter 4 attending explicitly to mood and movement and Chapter 5 drawing on loss, grief and horror, among others. Moreover, running through all five chapters so far is recognition of an evocation of empathy for the puppetised character, even if that is combined with shock and disgust at the actions they are compelled to undertake or if it is paradoxically associated with an evacuation of the sub-ject with which we are encouraged to empathise. Evidently, the generation of specific affective responses is central to these constructions of puppets and, affectively speaking, puppets can clearly be very powerful.

Significantly, though, this affective power was not confined to circum-stances in which the realism of the puppet was most convincing but – and most notably in Chapter 5 – was most acute at precisely those moments when the artifice was dramatically revealed. While this is seemingly at odds with emphasis among the theatre critics cited on realism in puppet performance, this counterintuitive power of the revelation of artifice has previously been acknowledged in puppetry literature as potentially the most powerful part of the drama (Gross 1997). It also lies at the heart of the paradoxical belief in the illusion despite knowing that it is an illusion, which holds particular rel-evance for non-representational geography, and specifically its conceptualisa-tion of the relationship between the cognitive (representational) and the affective (non-representational).

Non-representational geography is not about denying the importance of representations as vehicles of knowledge but about acknowledging that there are other forms of knowledge beyond representations (such as practice-based understandings), that representations are not neutral, and that repre-sentations are both the outcome of practical work and do work in the world, such that we need to think beyond the representation as an object to understand representation as a situated and constructive practice that is itself more-than-representational, for example, it is often affectively charged (McCormack 2005). Affect is an important concern within non-representational geography and has been variously defined as a sense of push in the world, a pre-personal force, a sense of connectedness or an intensity (Thrift 2004; Anderson 2006; Blackman 2012). Running through such diversity are emphases on affect as a felt intensity that occurs at a pre-reflective level, that connects us with other people, objects and events and that can be recognised either on an individual level (such as a sense of trepidation or excitement) or a collective level (such as an atmosphere of joy or grief). In the current context, it is the relationship between the representational (cognitive, conceptual knowledge) and the non-representational (felt, bodily sensations) that is most important, as a central feature of affect in geography is that – as pre-reflective – it is deemed cognitively ungraspable (Gibbs 2009). Fundamentally, while our pre-reflective/affective experiences influence our behaviours and cognitions (as when fear makes us more alert) we are unable to reach into our affective experience and bring it to formal conceptual cognition without losing something in the process because they operate according to different logics (McCormack 2006). Previously, I have suggested an alternative perspective on this key relationship between the cognitive and the pre-reflective from beyond geography (Banfield 2016) and I will return to that perspective in due course, but my starting point for this discussion is to question how the nature of the puppet as uncanny fits within conventional geographical understandings of that relationship. In other words, I am interested in whether and how non-representational geography can currently accommodate or account for this disconnect identified in Chapter 5 between what we know cognitively and what we believe performatively or feel on a pre-reflective level when watching puppet performances.

On a cognitive level, we can seemingly be acutely aware of the artificiality of the puppet yet continue to be engulfed in the affective charge that it has generated, as in the case of the radical strengthening of the border by the withdrawal of the puppeteers from the horse puppet that counterintuitively accentuated the affective quality of the scene. We might consider this to be due to a willing suspension of disbelief or simply a persistence of affective power beyond the original source of that affective impact, but these seem somewhat inadequate as explanatory accounts. At stake here is the status of belief: Whether we consider believing to be a cognitive act or a pre-reflective disposition. Dictionary definitions accommodate both, as belief is defined as faith, trust, intuition or the conviction of the truth of something (Chambers 2011), where faith, trust and intuition have strong pre-reflective associations but conviction as the state of being convinced is more cognitively oriented as to convince is to persuade by evidence (Chambers 2011). We have a dilemma, then, in that we must still account for the willing persistence of belief in the puppet at the pre-reflective level despite the evidence to the contrary at the cognitive level. Non-representational geography's perspective that the pre-reflective can influence the cognitive but not vice versa becomes problematic when we try to explain how we can pro-actively maintain affective belief in full knowledge of its baselessness if we are unable to reach into our pre-reflective experience from our cognitive understanding. Similarly, if our pre-reflective experience influences our cognitive thinking but not the other way around, presumably the strength of our belief should mask the revelation of the artifice of the puppet, but that does not happen if we perceive and acknowledge that artifice. How can we account for the maintenance of belief alongside evident awareness of the illusion?

My thinking is that two aspects are at play here, each of which will be addressed in turn: (1) The relationship between the border and the borderscape and (2) the need for a more flexible and generous understanding of how the cognitive and the pre-reflective relate to each other. With respect to the first of these, my suggestion is that while we might be cognitively aware of the artifice at the level of the border/s as those borders have been made abundantly and unavoidably clear, we can maintain our pre-reflective belief in the totality and reality of the puppet at the level of the borderscape because the totality of the puppet is more-than-cognitive. We perceive the breakdown of the puppet–human assemblage, but as this is only one aspect of what constitutes the puppet across the borderscape, we feel the unity of the puppet despite this. Perhaps, then, the core duality at the heart of the uncanniness of the puppet is not that between inanimate object and living subject, or that between belief and disbelief, but that between cognitive and pre-reflective engagements with the puppet, which are – it seems – spatialised to the extent that the broader affectivity within the borderscape can sustain the unity and vitality of the puppet despite the cognitive knowledge of its dichotomous and inanimate nature at the border/s. This proposed relationship between the border/s and the borderscape potentially enables us to disbelieve cognitively but believe affectively. However, with respect to the second of these aspects, we still need to address the difficulty in accommodating this phenomenon within conventional understandings of how the cognitive and pre-reflective relate in non-representational geography. It is at this point that I return to an alternative perspective on this relationship as previously articulated in the context of artistic practice (Banfield 2016).

The philosophical and psycho-therapeutic work of Eugene Gendlin is informative here, as it allows both for the pre-reflective (which is termed 'the implicit') to function independently of cognition and for some capacity to reach into our pre-reflective experience cognitively. Gendlin's thinking and terminology are complex, so I do not engage here with a detailed exposition of that body of work (for a geographically oriented exposition, see Banfield 2016), keeping my comments at quite a high (overview) level and confining them to this specific relationship between the cognitive and pre-reflective. Gendlin proposed that although we do not commonly draw intentionally upon our pre-reflective experience (termed implicit understanding), we do have some capacity to do so and have become unaccustomed to doing so through reliance on linguistic convenience and convention (Gendlin 1993,

1997). Gendlin's psycho-therapeutic work was directed towards facilitating and developing this potential and capability to enable those who had been (for example) traumatised to bring into cognitive awareness aspects of that experience that feel inarticulable but need to be articulated for the purposes of treatment (Gendlin 2009). For Gendlin, our conceptual (explicit) language came from our implicit understanding, so not only do we have some (albeit rusty) capacity to engage intentionally and cognitively with our implicit (pre-reflective) understanding, but that implicit understanding came before explicit understanding and can thus function independently of it (Banfield 2016). On the one hand, this temporal primacy and independent functioning of implicit understanding allow for the disconnect between affective and cog-nitive understandings, such as that identified above. On the other hand, the emergence of explicit understanding from implicit understanding renders questionable the assumed inability to dig into our pre-reflective experience from a conceptual or cognitive perspective on the basis that they are incom-mensurable, potentially facilitating the willing maintenance of affective belief. We might lose something in translation from implicit to explicit but that does not mean that there is no potential to work cognitively with the implicit, at least to some degree. Gendlin's work, then, seemingly offers a way to accommodate the uncanniness specific to puppets that has long been rec-ognised, has been variously conceptualised and has been confirmed again in this volume.

Moreover, Gendlin's work provides a framework within which we can situ-ate the spatialisation of the breaking down of cognitive belief at the border while simultaneously maintaining (strengthening even) the affective belief across the borderscape, through his notion of felt sense. This felt sense is a pre-reflective comprehension of the entirety of a situation, phenomenon or experience, etc., but it is important to remember here that explicit understand-ing comes after implicit understanding, so although felt sense is pre-reflective, it does not preclude cognitive understanding as there are pre-reflective under-pinnings to conceptual/cognitive understandings. Implicit understanding, then, both precedes and underpins cognitive/conceptual/explicit understand-ing, while felt sense encompasses both implicit and explicit understandings because the explicit is grounded in (and sprang from) the implicit. We might think, then, that the borderscape is a spatialisation of felt sense in the context of puppets as it encompasses both implicit and explicit (cognitive and affec-tive) understandings of the puppet concerned. Significantly, though, because the implicit underpins the explicit and because the implicit can function inde-pendently of the explicit, the rupturing of the puppet at the explicit (cogni-tive) level (the border) need not upset the totality/vitality/believability of the puppet at the implicit (felt, pre-reflective) level (the borderscape).

On this reading, not only can the implicit (affective) understanding of the puppet as a continuing totality be sustained despite the breakdown of the explicit (cognitive) understanding of the puppet, but we can also appreciate how this happens. The cognitive (explicit) understanding is grounded in, sprang from and is encompassed by the pre-reflective (implicit) understanding of the puppet, such that seemingly incompatible elements can be understood in an implicit fashion even though they do not make any sense in an explicit or logical fashion. For Gendlin, conceptually contradictory ideas or statements can be understood because even though they do not make sense on an explicit (cognitive) level, they do make some sense on an implicit (pre-reflective) level. This Gendlinian perspective is thus informative with respect to both puppets and the grotesque. With puppets, it accommodates the maintenance of the implicit understanding of the puppet as a totality despite the explicit understanding contradicting this, and with the grotesque, it accommodates integration or connectivity on an implicit level of incongruities and contradictions at the explicit level.

Both the uncanniness of puppets and the grotesque that lies at their heart can seemingly be accommodated within this Gendlinian perspective by accounting for the acknowledgement of dichotomies even while sustaining belief in a broader totality that encompasses those dichotomies, because the belief and the knowledge operate according to different *yet connected* logics. Ultimately, this analysis suggests that we need to spatialise the notion of the puppet as uncanny with respect to the relationships between both the border and borderscape and the cognitive and affective, and that we need to be more welcoming towards theorisations that allow for more diverse and bidirectional relationships between the cognitive and the pre-reflective (affective) than currently seems to be the case in non-representational geography.

## Conclusion

Through the varied analyses in Part 2, then, this conclusion delivers two primary contributions: (1) A revised dimensional framework that incorporates moral messaging as a purposeful use of puppets, borderscaping as a relational practice of puppet construction and grotesquery as a qualitative characteristic of puppetness and (2) a spatialisation of the uncanny with respect to both the borderscape and the relationship between the cognitive and the affective that encourages further disciplinary engagement with the philosophical and psycho-therapeutic work of Eugene Gendlin.

Further, though, borderscaping has been formulated as a practice by which puppets are constructed and a continuum of theatrical–radical dissolution or strengthening of the border has been conceived as a suite of strategies for utilising the puppet–human border in constructing a puppet. A diverse array of bodies grotesque and characters contorted have been identified – most notably in the simultaneous hyper-puppetisation as an archetype and de-individualisation as a character of Mr Punch – and the character contorted has been reinforced as a refinement to understandings of the grotesque to accommodate more psychologised formulations of the grotesque. The significance of specific puppet forms or types to the potential constructions they can generate has also been highlighted, and through this, faciality and pedality have been identified as important factors in configuring human–puppet relations. Moreover, it was suggested that the borderscape might help us to accommodate different perspectives on the uncanniness of puppets, affect was highlighted as a pervasive undercurrent running through these chapters, and the affective power counterintuitively generated by revealing the artifice of the puppet was explored by bringing together notions of the uncanny, the borderscape and the cognitive-affective relationship.

Finally, an answer to the question of what grants puppets their performativity has been generated, which is summarised in the following four aspects: Explicit interrogation of puppetness, the borderscape as an active arena for this interrogation, different approaches to this practice of borderscaping (theatrical–radical dissolution or strengthening of the border/s at its heart) and the relationship between implicit and explicit understandings of the puppet that allow for the continuance of belief in the puppet despite being fully aware of its artificiality.

## References

Anderson, B. (2006) Becoming and being hopeful: Towards a theory of affect. *Environment and Planning D: Society and Space*, 24: 733–752.

Banfield, J. (2016) *Geography meets Gendlin: An exploration of disciplinary potential through artistic practice*. Palgrave Macmillan: New York.

Banfield, J. (2020) 'That's the way to do it!': Establishing the peculiar geographies of puppetry. *Cultural Geographies*, 28(1): 141–156. doi:10.1177/1474474020956255.

Blackman, L. (2012) *Immaterial bodies: Affect, embodiment, mediation*. SAGE: London.

Chambers. (2011) *The chambers dictionary*, 12th Edition. Chambers Harrap: London.

dell'agnese, E. & Amilhat Szary, A.L. (2015) Borderscapes: From border landscapes to border aesthetics. *Geopolitics*, 20: 4–13.

Dittmer, J. (2010) Comic book visualities: A methodological manifesto on geography, montage and narration. *Transactions of the Institute of British Geographers*, 35: 222–236.

Dittmer, J. & Latham, A. (2014) The rut and the gutter: Space and time in graphic narrative. *Cultural Geographies*, 22: 427–444.

Gendlin, E.T. (1993) Words can say how they work. In: R.P. Crease (ed) *Proceedings, Heidegger conference*. State University of New York: Stony Brook, 29–35.

Gendlin, E.T. (1997) The responsive order: A new empiricism. *Man and World*, 30: 383–411.

Gendlin, E.T. (2009) We can think with the implicit, as well as with fully-formed concepts. In: K. Leidlmair (ed) *After cognitivism: A reassessment of cognitive science and philosophy*. Springer: London, New York, 147–161.

Gibbs, A. (2009) After affect: Sympathy, synchrony and mimetic communication. In: M. Greg & G.J. Seigworth (eds) *The affect theory reader*. Duke University Press: Durham, NC, 186–205.

Gross, K. (1997) Love among the puppets. *Raritan*, Summer 1997(17): 67–82.

van Houtum, H. & Eker, M. (2015) Borderscapes: Redesigning the borderline. *Territorio*, 72: 101–108.

McCormack, D. (2006) For the love of pipes and cables. A response to Deborah Thien. *Area*, 38: 330–332.

McCormack, D.P. (2005) Diagramming practice and performance. *Environment and Planning D: Society and Space*, 23: 119–147.

Mello, A. (2016) Transembodiment: Embodied practice in puppet and material performance. *Performance Research*, 21(5): 49–58.

Thrift, N. (2004) Intensities of feeling: Towards a spatial politics of affect. *Geografiska Annaler, Series B: Human Geography*, 86: 57–78.

# Part 3
# Transformational puppets

# Introduction to Part 3

My intention in this part is to engage explicitly with the transformation of puppets into humans and humans into puppets, to interrogate in greater detail the varied ways in which puppets and humans are mutually incorporated. In part, this picks up on the recognition that a key aspect of puppet theatre is its talent for metamorphosis in which everything has the propensity to become something other than it originally was (Gross 2011). Combining contradictory terms, puppets embody a dynamic connection between opposites, including subject and object, living and non-living (Cappelletto 2011; Jurkowski 2013), in a transformative assemblage of disparate elements (Buchan 2011). In part, though, this emphasis on transformation also picks up on contemporary disciplinary concerns for becoming rather than being. With the rise of assemblage thinking and relational ontologies has come a reworking of the bases of phenomenology to shift the emphasis from the essences of things to the emergence of what appears as essences and growing appreciation of the intertwining of humans and nonhumans and the distributed nature of subjectivity and agency (Thrift et al. 2010). The metamorphic propensity of puppets, then, is ideally suited for geographical analysis given the current post-phenomenological and posthuman emphases within the discipline, especially in cultural geography.

Becoming assumes that the world is dynamic and open-ended, drawing on strands of thinking from Spinoza, Bergson, Heidegger and Merleau-Ponty among others, and – consequent to this – neither entities nor space are fixed as both are considered processual (McCormack 2009), drawing attention to both the transformative potential of entities and the emergent nature of space. Whether considered individually, collectively or both, the inherent indeterminacy of entities and their inter-relations affords the be/coming together of subjects through encounters among an open multiplicity of human and nonhuman entanglements (Connolly 2011; Simpson 2017; Brice 2020; Ghoddousi and Page 2020). Much geographical work on becoming draws heavily on the writings of Deleuze and Guattari (see, for example, Bonta 2005, 2010; Blanco et al. 2015), especially their notion of becoming-other, whereby new entities and subjectivities emerge from rhizomatic inter-linkages among shifting assemblages without a predetermined endpoint

DOI: 10.4324/9781003214861-14

(Bonta 2005). Consistent with this, in Chapters 6 and 7, I explore becoming-human and becoming-puppet, respectively, before integrating the findings of these two chapters in the Conclusion to Part 3 by developing a composite understanding of 'puppet-becomings', which refers to transformations in both directions.

**Chapter summary**

The question guiding this third part is: How are puppets constructed? While we already have a partial answer to this question, for example, through the idea and practices of borderscaping, the incorporation of contradictory forms and intercorporation, my interest in this part is in deepening these understandings. Specifically, I adopt a tighter focus on core factors in the construction of puppets and their worlds that emerged from Part 1 – materiality, subjectivity and sociality – as distinct approaches to intercorporation. In other words, I explore how puppets become humans (Chapter 6) and how humans become puppets (Chapter 7) through varied intersplicings of material, subjective and social factors.

In Chapter 6, I focus solely on Carlo Collodi's mischievous puppet Pinocchio as the archetypal puppet who wants to be a real boy and who undergoes a series of metamorphoses. Through a close analysis of four illustrated literary versions of this tale, with emphasis on the visual content of the works, the conveyance of both the general construction of Pinocchio (focusing on his materiality and sociality) and the specific means of crafting his metamorphoses is established, thereby constituting his subjectivity. Along the way, implications of specific constructive approaches for the affective power of the imagery and the integrity of the tale are explored, and the contradictions that are brought into constructive union are used to elaborate on Collodi's own understanding of the contrariness of the puppet. Considering these constructions in spatial terms brings to light a different perspective on how Pinocchio is constructed, not only through the contradictory constructions themselves but also through the varied spatialities that they generate. Specifically, the possibility of incongruence between the underpinning borderscape and the spatialities that emerge from it is highlighted, which can undermine both Pinocchio's ambiguous status and the integrity of the work.

Chapter 7 attends to transformations of humans into puppets in 14 cultural works. Initially, I review these works in relation to the understandings of puppets that have emerged through the preceding chapters, generating additional features of puppet construction and puppet-becoming, before developing a 'patchwork' schematic of different perspectives that the works adopt towards their progressive transformations, each of which has its own spatiality. Through these discussions, a set of constructive complementarities are identified that supplement constructive contradictions in the constructions of human-to-puppet transformations. Within these perspectives, I also identify ten specific devices that are employed and use these to conceptualise a suite of corporealities, providing both more detail and a broader context for the foregoing emphasis on intercorporeality, that work either in isolation or in combination to configure these diverse becomings. This – in turn – is rearticulated as a series of ecologies and technologies that prioritise human-oriented and puppet-oriented features, respectively, which facilitate the integration of the multiple conceptualisations that have emerged through this volume to forge a fuller and more formalised description of the borderscape and borderscaping.

## Conclusion

Part 3, then, moves beyond consideration of the construction of 'puppet' as a state of being to explore the construction of puppet transformations or becomings, whether from puppets to humans or the other way around, to tease out in its two chapters specific ways in which these transformations are accomplished. Subsequently, the Conclusion to Part 3 draws these two sets of analyses together, both in comparative terms and through integrating their respective findings. This affirms the applicability of each set of findings beyond its own dataset and the translatability of specific aspects of the findings to accommodate particular puppet-becomings. It is also used to schematise these combined and interwoven ecologies and technologies of puppet-becoming as a constructional framework that supplements the cumulative dimensional framework presented in the Conclusion to Part 2. Finally, the primary contributions of Part 3 to geography are outlined, picking up on the unearthing of literary micro-spatialities from Chapter 6 to propose novel approaches to geocriticism, and on the constructional devices identified in Chapter 7 to contribute to understandings of corporeality and debates about a posthuman humanism.

## References

Blanco, G., Arce, A. & Fisher, E. (2015) Becoming a region, becoming global, becoming imperceptible: Territorialising salmon in Chilean Patagonia. *Journal of Rural Studies*, 42: 179–190.

Bonta, M. (2005) Becoming-forest, becoming-local: Transformations of a protected area in Honduras. *Geoforum*, 36(1): 95–112.

Bonta, M. (2010) Ornithophilia: Thoughts on geography in birding. *The Geographical Review*, 100(2): 139–151.

Brice, S. (2020) Geographies of vulnerability: Mapping transindividual geometries of identity and resistance. *Transactions of the Institute of British Geographies*, 45(3): 664–677.

Buchan, S. (2011) *The Quay Brothers: Into a metaphysical playroom*. University of Minnesota Press: Minneapolis, MN and London.

Cappelletto, C. (2011) The puppet's paradox: An organic prosthesis. *RES: Anthropology and Aesthetics*, 59/60: 325–336.

Connolly, W.E. (2011) *A world of becoming*. Duke University Press: Durham, NC.

Ghoddousi, P. & Page, S. (2020) Using ethnography and assemblage theory in political geography. *Geography Compass*, 14(10): e12533.

Gross, K. (2011) *Puppet: An essay on uncanny life*. University of Chicago Press: Chicago, IL.

Jurkowski, H. (2013) *Aspects of puppet theatre*, 2nd Edition. Palgrave Macmillan: Basingstoke.

McCormack, D.P. (2009) Becoming. In: R. Kitchin & N. Thrift (eds) *International encyclopedia of human geography*. Elsevier Science Ltd: Amsterdam and London, 277–281.

Simpson, P. (2017) Spacing the subject: Think subjectivity after non-representational theory. *Geography Compass*, 11: e12347.

Thrift, N., Harrison, P. & Anderson, B. (2010) "The 27th Letter": An interview with Nigel Thrift. In: B. Anderson & P. Harrison (eds) *Taking place: Non-representational theories and geography*. Ashgate Publishing Ltd: Farnham and Burlington, 183–198.

# 6 Puppets becoming human

In this chapter, I focus solely on Pinocchio as the archetypal puppet who wants to – and in most versions does – become human. As a symbol of modernity, Pinocchio reflects a nineteenth-century cultural interest in the fusing of humans and machines (Lucas 2011; Pizzi 2011). Clearly exhibiting the contrariness already established as characteristic of puppets and the grotesque, the tale is constructed upon a series of contradictions, such as between fantasy and reality, animate and inanimate, renewal and growth (Perrot 2011; Riva 2011; Wilson 2011). This essential ambiguity has also been considered in relation to parallels between narrative and illustrative aspects of the Pinocchio story, which in their juxtaposition are proposed to compose an unresolved totality that mirrors the ontological ambiguity of the titular puppet (Wilson 2011). Picking up on this narrative–illustrative ambiguity, this analysis draws upon four illustrated written versions of the tale, but as the visual content is more varied than the verbal content across these works, the primary analytical emphasis is on the visual representation of Pinocchio. The works analysed are:

- Collodi, C. (1862/2014) *Pinocchio*. Collector's Library: London.
- Collodi, C. (1862/2011) *Pinocchio: The Story of a Puppet*. The Folio Society: London.
- Daly, A. (2014) *Pinocchio*. Ladybird Books Limited: London.
- Morpurgo, M. (2015) *Pinocchio by Pinocchio*. Harper Collins: London.

Compared to the original Collodi text, the Daly text is a Ladybird version for very young children, and the Morpurgo text is aimed at older children and is written from Pinocchio's own perspective. For the sake of clarity, the two Collodi texts are identified by their respective publishers, abbreviated to Collector's and Folio. The other texts are cited by the author, with all references to them italicised.

In addition to reviewing the general visual style of each work, the visual analysis focused on:

1. The cover design.
2. The visual construction of Pinocchio.

DOI: 10.4324/9781003214861-15

3.  Pinocchio's four metamorphoses:
    a.  From log to puppet.
    b.  From puppet to donkey.
    c.  From donkey back to puppet.
    d.  From puppet to real boy.

4.  Three other flagship events in the tale that highlight Pinocchio's puppet status:

    a.  Burning his feet in the fire.
    b.  His experience at the puppet theatre (threat of being thrown onto a fire).
    c.  Having his nose pecked back down to size by woodpeckers.

This generated a data set of 56 images, covering approximately 50% of the visual content of the *Folio* and *Daly* versions as these contained relatively few images and approximately 25% of the *Collector's* and *Morpurgo* versions as these are more thoroughly illustrated, although many of these illustrations are tokenistic rather than specific to the tale. The overall visual style of each work differs, with all the images in the *Collector's* version being black and white with many in silhouette, thereby reducing the amount of detail that can be conveyed, whereas those in the *Folio* version are fewer, exquisitely detailed and mostly vibrantly coloured. The *Daly* version conforms to the corporate style of the Ladybird series, with one full-page image opposite concise text, bright and blocky colours and simplified images. By contrast, the images in the *Morpurgo* version are in variegated greyscale, varied in size and normally detailed. Beyond these differences, though, each version constructs Pinocchio very differently, and these differences form the focus for this analysis.

This chapter is organised in five parts, the first three of which constitute the empirical analysis and the latter two of which work through its implications. The analysis adopts materiality and sociality as orienting themes, as these have emerged as core aspects of puppet constructions in Part 1 and have been consistent emphases through the subsequent chapters. Initially, I address the construction of Pinocchio's materiality, drawing both on a study of his general visualised representation in each work and one key event – Pinocchio burning his feet – that primarily speaks to Pinocchio's materiality but also introduces the consideration of his sociality. Subsequently, I address sociality directly, again by considering both generalised depictions and the representation of one specific event – Pinocchio's encounter with the Fire-Eater at the puppet theatre – that is of primary significance to sociality but also incorporates materiality. Thirdly, I analyse the visual depictions of Pinocchio's ontological metamorphoses as explicit examples of Pinocchio's becoming, and these analyses progressively uncover inconsistencies in Pinocchio's construction within individual works, incongruities between his narrative and illustrative construction within individual works and contradictions in the construction of his becoming between works.

Subsequently, the implications of these analyses are explored. I begin by developing Carlo Collodi's own articulation of puppetness as the 'but' of the puppet, drawing upon the third key event – involving the woodpeckers – to test a conceptual schematic of the eight constructive contradictions – or topological gulfs – identified in the earlier analysis. Finally, I focus on ways in which the *Morpurgo* version of the tale diverges from the others, in relation to Pinocchio's materiality, sociality and metamorphoses, to consider the implications of these divergences for the borderscape and the construction of the puppet that is (dis)allowed by it. Through this discussion, I also explore both the relationship between the borderscapes that underpin these works and the spatialities that emerge from them, and the additional role played by these spatialities in the construction and becoming of Pinocchio.

## Pinocchio's materiality

The analytical summary provided in Table 6.1 establishes that while none of the works excludes any suggestion of woodenness, the *Folio* and *Collector's* versions give a stronger sense of Pinocchio's wooden nature in the clear visibility of joint mechanisms and grain in his complexion, whereas the *Daly* and *Morpurgo* texts tend to depict Pinocchio as human. While there is some evidence of grain on Pinocchio's limbs in the *Daly* version, he is also portrayed here with rosy cheeks and cherry red lips, while in *Morpurgo*'s autobiographical tale Pinocchio is depicted as fleshy rather than wooden and with cartoon-style eyes. Similarly, there is some suggestion of joint mechanisms in these works, but this is as folds or simple lines without any sense of separation between – for example – shin and thigh. This highlights a distinction between the treatment of Pinocchio's face and body, with any signs of woodenness in *Daly* and *Morpurgo* being confined to the body, maintaining a facial appearance of fleshy humanity. While a similar distinction is evident in the *Collector's* version, the *Folio* version displays grain on Pinocchio's face and nose as well as his body. Pinocchio's all-important nose also varies, with differences in naturalism of scale and form, but the *Morpurgo* version is most notable as the only one in which Pinocchio's nose is permanently extended, effectively erasing in visible terms a key aspect of the tale.

Overall, Pinocchio is constructed materially as a mixture of human and puppet in different ways:

- In the *Folio* version, he is a wooden figure with episodic fleshiness.
- In the *Daly* version, he is childlike and cherubic in face and body but with mature limbs and is fleshy in the face and wooden in the body.
- In the *Collector's* version, he is a wooden figure with a humanly expressive face and localised fleshiness.
- In the *Morpurgo* version, he is a cartoonised human child with a long nose.

*Table 6.1* Summary of Pinocchio's materiality

| Version | Folio | Ladybird | Collector's | Morpurgo |
|---|---|---|---|---|
| **General** | Bodily proportions seem quite adult, although neck slightly thin. Joint mechanisms shown clearly. Grain consistent, including nose. Complexion brown. Nose human-scale but not realistic. When long, nose tapers very gently. Hands sometimes wooden and stiff, but at others realistic (e.g., fleshy base of thumb). Never depicted running. Fairly inactive: sitting, hanging, riding. *Overall*: a wooden figure with human capabilities and episodic fleshiness. | Composite: a pear-shaped body with slender limbs, a very thin neck and a round face. Rosy cheeks, cherry red lips; grain on limbs but complexion on face. Nose like a cork, and a different colour to the face. When long, nose tapers very gently. Joints indicated but shown as creases/folds. Some images show him walking and running. *Overall*: Childlike and cherubic in face and body but with mature limbs. Fleshy and wooden. | Bodily proportions seem adult. A cylindrical body. Grain on body and limbs. Face humanoid compared to body and limbs: rounded features Nose naturalistic in scale and appearance. When long, nose tapers very gently. Joints suggested, not in detail. Hands sometimes fixed/static but also showing a range of movements Several active scenes: running and fighting. *Overall*: a wooden figure with a humanly expressive face. Localised fleshiness. | Bodily proportions of a toddler – chubby. No grain, so seems fleshy. Joints indicated with lines but no sense of joint separation. Cartoon style eyes. A conical nose, permanently over-sized Numerous active scenes: walking, playing, running and dancing *Overall*: a cartoonised human child with a large nose. Fleshy. |

| Burning feet | Pinocchio on the floor with lower legs charred and still aflame, crying and with his arms outstretched to Geppetto, whose own posture mirrors this stance, protruding into room through window. This gestural reciprocity reflects their relationship and establishes it as father–son. A complex mix of contradiction and mutuality that speaks to themes of the narrative, for example, fire showing the apparently benign as potentially malign. The image encapsulates the ethos of the work in its representational content and its aesthetic style. Complex interplay. | Pinocchio lies on the floor, with his feet charred but not aflame, while Geppetto climbs through the window. No obvious sign of concern, and no gestural reciprocity in a highly sanitised scene that masks the relationship and precludes any mutuality between the identities of Pinocchio and Geppetto. The humanised context for Pinocchio's unfortunate situation underplays the ambiguity at the heart of the tale. The significance and meaning of this episode are also simplified and sanitised by confining it to a human context. | Two images: 1) Pinocchio asleep on a wooden chair with his feet resting on the lit brazier and 2) Pinocchio on his knees with his lower legs charred, tapered and lacking feet. Pinocchio has his head in his hands and is crying while a cat plays with wood shavings behind him. These images establish Pinocchio's membership of a material world rather than a human world. Although the cat is evident, its disinterest in Pinocchio's plight nullifies any animal sociality, reinforcing the materiality of Pinocchio. A strong material emphasis with no human referents. | Two images: 1) Pinocchio curled up asleep with smoke rising from his feet and 2) Pinocchio being returned to Geppetto and wife, all with smiles on their faces and arms outstretched. Pinocchio's lack of distress emphasises his familial relations rather than personal capacities and underplays his materiality. There is three-way gestural reciprocity between Pinocchio, Geppetto and his wife. Human context only: A conventional nuclear family. |

Pinocchio burning his feet while asleep is the first time that he comes to physical harm. Two of these works (*Daly* and *Morpurgo*) visually confine the event to a human social context, one (*Collector's*) emphasises his woodenness as a puppet and the last (*Folio*) delivers a complex interplay of materiality, subjectivity and relationality. While the *Daly* version depicts the basic aspects of this event, there is no visual indication of the relation between Pinocchio and Geppetto, nor of concern on the part of either, precluding any subjectivity for Pinocchio and Geppetto, underplaying the ambiguity in Pinocchio's being and resulting in a contradictory outcome of a humanised Pinocchio in form but without any human affective or relational characteristics. In the *Morpurgo* version, Pinocchio's joy at being reunited with his parents – while instilling subjectivity in the puppet through the gestural reciprocity of their outstretched arms – emphasises his human familial relations, thereby underplaying his materiality.

The *Collector's* version shows Pinocchio on his knees with his lower legs charred, tapered and lacking feet and with his head in his hands crying while a cat plays with wood shavings behind him, establishing Pinocchio's membership of a material world rather than a human world but with human emotionality. The cat is interesting here, though, as – although Pinocchio has numerous animal companions (see below for more on this) – the disinterest of this cat in Pinocchio's plight nullifies any animal sociality in the broader narrative, thereby reinforcing the objectivity of Pinocchio. In this version, there is a strong material emphasis with no human referents, yet Pinocchio exhibits the emotionality of humans in a reversal of the contradiction in *Daly's* version.

The *Folio* version lies somewhere between these two extremes. The gestural reciprocity between Pinocchio and Geppetto reflects their affective bond and establishes their relationship as father and son, contrasting starkly with Pinocchio's evident wooden materiality, which is reinforced by his lying prostrate on a wooden floor against which Pinocchio is partially camouflaged. This image presents a complex mix of contradiction and mutuality between Pinocchio and Geppetto, such as father/son, protector/vulnerable, as well as Pinocchio's own internal contradictions, that speaks to core themes of the narrative and encapsulates the spirit of the Pinocchio tale in both its representational content and its aesthetic style.

Collectively, Pinocchio's materiality is constructed through a series of contradictions, with a generalised emphasis on Pinocchio's form consisting of a fleshy head on a wooden body, although in *Morpurgo*'s tale he is distinctly and consistently human. He is sometimes constructed as human in form but object-like in lacking human relations and emotions (*Daly*) but equally and conversely as material in form but with human emotions (*Collector's*), drawing a further distinction between bodily substance and psychological capacities, while the *Folio* version delivers a more complex construction. These works, then, reveal at least four ways in which the contradictions between human and puppet are operationalised – flesh/wood, form/expression, body/head, body/mind – and the varied ways in which these contradictions are constituted reflect Pinocchio's ambiguous ontological status between as much as within works.

## Pinocchio's sociality

The analytical summary in Table 6.2 provides new perspectives on Pinocchio's sociality as he prompts us to think beyond the relationship between puppet and human due to his many and varied relations with animals, including a fox, a cat, chickens, weasels, dogs and fish, to name a few. Consistent with the Italian fable tradition (Consolo 2011a), this prevalence of animal associations and comparisons is evident in all works except the *Daly* version. In the *Folio* version, animals are also used for scaling effects to establish Pinocchio's small stature and emphasise his vulnerability, while in *Morpurgo*, they are often used to break up the text or maintain a sense of visual flow across pages with no text, establishing a highly generalised animal sociality.

With the exception of Geppetto and the Fairy, humanoid characters are predominantly nasty to Pinocchio, caging him like a dog, whipping him as a donkey, or frying him like a fish, many of which are depicted visually according to the overall visual style and narrative emphasis of each work. With the exception of the dastardly cat and fox, animal characters are more generous towards Pinocchio, warning him of the likely consequences of his behaviour and rescuing him from scrapes. While this is not a clear or complete distinction, as the Fire-Eater eventually gives Pinocchio money for Geppetto despite threatening to burn Pinocchio and Pinocchio is both chased and saved by a dog, on the whole Pinocchio is more positively associated with animals than humans.

This is reinforced in relation to violence in the tale. In the two Collodi versions, animals are depicted as animals and the human use of animals and other humans is equally brutish, but the violence depicted (especially visually) is human to human rather than human to animal. In the *Collector's* version, for example, human–human violence is depicted among both adults (p11) and children (p134), human–animal maltreatment is shown in relation to Pinocchio's time as a donkey in the circus (p185), and the equivalence drawn between the two is evident when Pinocchio is treated as if he was a dog when he is clearly a puppet-person (p103). In *Morpurgo*'s version, however, animal characters are strongly anthropomorphised and the violence receiving the greatest attention is human to animal, as the human–human violence is largely downplayed but there are four separate images of the maltreatment of donkeys (pp213, 234, 243, 263). In Collodi's original tale, Pinocchio is mistreated whether he is deemed puppet, animal or human but in *Morpurgo*'s version, Pinocchio's suffering is emphasised by his animal associations. While the corporeal vulnerability of animals enhances our empathy for Pinocchio given their comparatively more inhumane treatment, it also simultaneously reconfirms his existence as a living being rather than a material object, underplaying his essential ambiguity. Consequently, dichotomous assumptions about Pinocchio's status between puppet and human, or toy and boy, are

*Table 6.2* Summary of Pinocchio's sociality

| Version | Folio | Ladybird | Collector's | Morpurgo |
|---|---|---|---|---|
| **General** | Few images include human(oid) characters. Sense of community and solidarity with other puppets at theatre. Animals emphasise Pinocchio's small size: equal to fish in pan, dwarfed by snake and dogfish. Only core animals included. Images focus on key events. Mix of realistic and slightly surreal imagery. Mostly brightly coloured but occasionally dark or black/white. Inside dogfish shown. | More humans depicted than animals. Puppet theatre well populated with puppets. Only core animals included (e.g., no snail, dog, parrot or gorilla) No significant scaling effects, e.g., with animals. Inside dogfish not shown. | Diversity of human characters: guards, children, fisherman. Fairy under-represented (only two small icons). Only one other puppet shown. Sometimes used as chapter markers (e.g., cat, fox, bird). Sometimes used to indicate key events (woodpeckers). Some direct contrasts between silhouette and detailed line images. Inside dogfish shown but fish itself not evident. | Mix of human and animal characters. Puppet characters humanised. Many animals with cartoon eyes; anthropomorphised. Sometimes used to break up text. Mixture of realistic and stylised imagery. More animals shown than are implicated in the tale. Inside dogfish shown. |

| Puppet theatre | The Fire-Eater's leather whip with snakes and fox tails at the end objectifies Pinocchio, while the potential for the puppets to be immobilised by the same material (rope) that would normally enliven them reinforces his ambiguous status. The peculiar combination of perspective and lack of perspective, realistic and unrealistic, 2D and 3D imagery, and light and dark reinforces the juxtaposition of puppet and human societies. While this disconcerting quality is powerful, it is not easy to discern quite what it is that generates it. | This image covers separate sub-events (transference of threat from Pinocchio to Harlequin and Pinocchio's offer to sacrifice himself to save Harlequin), making it difficult to decipher beyond a generalised sense of threat and puppet worry. While the threat is serious, it is both trivialised (by the Fire-Eater not being depicted as fierce in contrast to his description in the text) and generalised (by the uncertainty as to which puppet is the cause of greatest concern). Contradiction with text, conflation of events and confusion of threat. | Two images: 1) The Fire-Eater in a chair and 2) Pinocchio kneeling in front of the Fire-Eater, with his arms reaching up to him imploringly. In the background, Harlequin is held by two gendarmes with a look of alarm on his face. In the foreground is a fire-grate. Pinocchio is begging the Fire-Eater not to throw Harlequin on the fire, but the affective power is undermined by both the wizard-like appearance of the Fire-Eater and the lack of flames in the fire-grate, which downplays the danger. Understated in affective power. | Four images: 1) a carnival scene, with Pinocchio carried aloft in celebration, 2) the Fire-Eater holds Pinocchio up to his face by the scruff of his neck, with his mouth open in a roar, displaying shark-like teeth, 3) the Fire-Eater holds Pinocchio by the scruff of his neck at waist-height as he tries to stifle a sneeze and 4) a party scene, with Pinocchio being swung through the air in joy. The threat is downplayed (no sign of fire) in favour of the social and celebratory atmosphere. Sanitised threat, tranquilised terror and celebratory emphasis. |

overly simplistic, overlooking the additional animal nuance that features in both Collodi's original tale and *Morpurgo*'s retelling, but which really comes to the fore by comparing the two, and which can either emphasise or undermine his ambiguous status.

In relation to Pinocchio's time at the puppet theatre, these works also variously establish Pinocchio as a member of a puppet, animal or human community, with implications for the affective power of this threatening event. In the *Collector's* version, Pinocchio is begging the Fire-Eater not to throw Harlequin on the fire, clearly establishing his new sociality with the puppet community but *Morpurgo*'s version confuses Pinocchio's sociality in its depiction of this event, which is presented through four images. The presence of both human and animal characters in images one and four positions Pinocchio at the junction between these two social realms, but his status as a puppet – although declared in the text – is undermined by the depiction of the characters of Punch, Judy and Harlequin in humanised form. While this is consistent with the visual approach of this work and Pinocchio's offer in the narrative to sacrifice himself for Punch and Judy confirms a sense of solidarity, in the visuals the significance of Pinocchio's materiality is lost and his sociality is confused. The contrast in sociality between these works is also evident in their respective narratives, as in the original tale the puppets recognise Pinocchio and call themselves his 'wooden brothers' (*Collector's*, p43), whereas in *Morpurgo*'s version, they do not recognise Pinocchio despite Pinocchio describing himself as the most famous puppet the world has ever known. However, despite these differences, both works undermine the affective power of these events. In both works, the Fire-Eater is depicted as a wizard-like character who does not look fierce, while in the *Collector's* version the fire-grate into which Harlequin risks being thrown lacks a fire and the *Morpurgo* version focuses on the celebratory atmosphere, all of which mollify the sense of threat posed to the puppets. While this reflects the emphasis on illusion and the ambiguity between reality and fantasy in the original tale (Wilson 2011), for me it detrimentally impacts the affective force of the visuals and – through their relation to the narrative – the narrative itself.

The visual representation of this event in the *Daly* version is more affectively powerful than these works but loses some of its impact through its own conflation of the events covered and confused depiction of Pinocchio's puppet sociality. This image covers two sub-events, so it is difficult to decipher what is being shown beyond a generalised sense of threat and puppet worry. While Harlequin is most immediately in danger, the other puppets seem more concerned about Pinocchio, and the representation of only some of the puppet characters with strings further obscures his materiality and sociality, reinforcing both the interwoven nature of Pinocchio's materiality and sociality and the significance of consistency between the two. While there is affective power in this image, the lack of clarity as to the character/s with which the reader is intended to empathise means that this affect lacks a target.

In the *Folio* version, Pinocchio, Punch and Harlequin – without strings but clearly not fleshy humans – perform together in front of the fierce-looking Fire-Eater who looms over them threateningly. The depiction of ropes (top left) and whip (top right) further suggest threat, especially as they forge links to Pinocchio's other adventures (e.g., being hung from a tree, whipped as a donkey). While the whip with snakes and foxtails at the end could be seen to associate Pinocchio with animals, I interpret it as also objectifying both animals and Pinocchio, and this objectification of the puppets heightens the sense of threat despite there being no obvious immediate danger. The potential for the puppets to be immobilised by being bound with the same material (rope) that would normally enliven them reinforces both Pinocchio's objectified status and his new sociality with the obviously wooden puppets (established by jointed knees and rivets for knuckles). This image employs a peculiar combination of perspective and lack of perspective through a mixing of two- and three-dimensional imagery combined with a strong contrast between bold, bright colours and deep, dark shadows. Consistent with the juxtaposition of reality and fantasy that characterises the original work (Wilson 2011), this is both affectively powerful and significant in its encapsulation and communication of core themes, juxtaposing yet integrating very effectively the materialities and worlds of human and puppet.

Looking across these works, Pinocchio's sociality is sometimes established as strongly human, at others distinctly puppet, and yet others mostly animal, but these associations are not always consistent within a single work, most notably in *Morpurgo*, which denied Pinocchio's puppet sociality. The affective implications of this are notable, again especially so in *Morpurgo*, which significantly sanitised the sense of threat. In the *Collector's* version, the target of empathy was clear but the source of the threat was undermined, whereas in the *Daly* version, the source of threat was apparent but the character in greatest peril was ambiguous. While the *Folio* version also mollified the sense of a specific threat, the numerous visual references to the broader narrative multiplied the sources of threat discernible, while the foregrounding of the puppets established them as targets for empathy. Complementing the previous section, then, Pinocchio's sociality is constructed through a series of further contradictions on two dimensions: A social dimension, on which Pinocchio is variously and unstably constructed in relation to humans, animals and objects, and an affective dimension, on which a sense of threat might be discernible with/out the perceptibility of its source and on which empathy might be evoked with/out the identifiability of an appropriate target for it.

## Pinocchio's metamorphoses

Given the significance of Pinocchio's metamorphoses not only as central features of the narrative but in establishing him as a mythological as much as a literary figure (Consolo 2011b), it might be expected that every retelling of the tale would incorporate all four transformations, in visual as well as a

*Table 6.3* Topological Gulfs employed in visual constructions of Pinocchio

| | | | Folio | Daly | Collector's | Morpurgo |
|---|---|---|---|---|---|---|
| **Materiality** | 1 | Flesh/wood | ✓ | ✓ | ✓ | X |
| | 2 | Form/ expression | ✓ | ✓ | ✓ | X |
| | 3 | Body/head | X | ✓ | ✓ | X |
| | 4 | Body/mind | ✓ | X | ✓ | X |
| **Sociality** | 1 | Human | ✓ | ✓ | ✓ | ✓ |
| | | Animal | ✓ | X | ✓ | ✓ |
| | | Puppet | ✓ | ✓ | ✓ | X |
| | 2 | Source of threat clear? | Multiple | ✓ | X | X |
| | | Target of threat clear? | ✓ | X | ✓ | ✓ |
| **Whole work** | 1 | Ambiguity sustained? | ✓ | X | X | X |
| | 2 | Narrative consistency? | ✓ | ✓ | ✓ | X |
| **T1 (log–puppet)** | | Materiality | ✓ | N/A | ✓ | ✓ |
| | | Sociality | ✓ | N/A | ✓ | ✓ |
| **T4 (puppet–boy)** | | Materiality | ✓ | ✓ | ✓ | N/A |
| | | Sociality | Implied | ✓ | X | N/A |
| **Outcome** | | Transformed? | ✓ | ✓ | ✓ | X |

verbal form for illustrated works. However, Table 6.3 establishes that this is not the case. While the *Folio* version uses framing devices around the visuals for the transformations shown to highlight the significance of these events, none of the works visually presents all four metamorphoses, and – perhaps most notably – none of them shows the final transformation from puppet to boy in visual form, relying instead on an image of its outcome. This perhaps supports the assertion that Pinocchio's ambiguous status is essentially unpresentable (Wilson 2011), although the different works do achieve this in various ways and to differing degrees, as described above. Despite the patchy nature of the visual presentation of these metamorphoses, there are informative insights to be gleaned from their analysis, the most interesting of which emerge from transformations one (log to puppet) and four (puppet to real boy), which form the focus of this section.

The visualisation of the metamorphosis from log to puppet is most interesting in the *Morpurgo* and *Folio* versions, as it is not shown at all in *Daly* (relying on before and after images) and while the *Collector's* version does show the transformation, this is largely confined to a context of human social relations. The *Morpurgo* version tells of this transformation in three significant images. The first image extends Pinocchio's origins backwards in time to before his period as an inert log to his existence as part of a living, growing tree; the second draws out the contrast between his wooden materiality and his subjective potential as Geppetto is shown cradling a log with a face, and the third places him firmly within a human sociality, holding the hands of his 'parents'. In this version, the transformation is evident and a lineage of vitality is suggested from tree to puppet, but the imagery adopts a highly humanised focus on childlike cuteness. The *Folio* version contains one image of the in-between stage of Pinocchio's creation as part log and part puppet, in which Geppetto is depicted nose-to-nose with a half-made Pinocchio. Their facial features mirror each other, reflecting their relationship, while the expressivity of Pinocchio's face contrasts with the chisel in Geppetto's hand, establishing the ambiguity as to Pinocchio's status as object or subject. As Pinocchio-as-log lacks hair and his nose is already overly long, the relationship between the puppet and the creator is not based primarily on resemblance – although there is similarity in the shape of their mouths and the slope of their eyebrows – but on the mirror image between their respective faces. While the *Collector's* and *Morpurgo* versions also both foreground the father–son relationship between Geppetto and Pinocchio, the former through visual similarity and the latter through bodily gesture, the *Folio* version draws out the parallels between Geppetto and Pinocchio that run through the original tale: Pursuing dreams that turn out to have unanticipated consequences, regret at actions taken, and the affection that runs both ways. This image both evidences the transformation and encapsulates the tale, establishing the contradictions at the heart of Pinocchio's ambiguous status and fabulist adventures.

It is clearly possible to depict a part-puppet, part-human character, as some works manage to do so for Pinocchio's first transformation, yet none of the works considered here does so in relation to his final transformation (puppet to real boy), and the reason for this varies across the works. In the *Daly* version, the visual omission of the previous transformation (donkey back to puppet) determined the omission of this transformation, and the final image is of Pinocchio as a real boy, skipping merrily along a beach, watched by Geppetto and the Fairy. While this confirms the transformation and evokes the vitality with which Pinocchio is now imbued, it resolves rather than sustains the tale's core ambiguity.

Both the *Folio* and *Collector's* versions also visualise the outcome of this final transformation. In the *Folio* version, this takes the form of a framed portrait of Pinocchio, looking straight out at the reader with sparkly blue eyes and rosy cheeks. This is a fully humanised version of the cover image, but this real boy is strangely immobilised in a portrait form, complete with a picture frame. Conversely sustaining Pinocchio's core ambiguity even while resolving it, this image is contrary, effective and entirely apt. The final image in the *Collector's* version is a small picture of Pinocchio as a real boy with more naturalistic eyes and hair. This is also a head-only image of Pinocchio, who – lacking any means of movement – is similarly immobilised despite having become real. However, the tale's core ambiguity is not sustained as the lack of context or frame for the head-only image means that he is not re-objectified as he was by being turned into a portrait.

In the *Morpurgo* version, there is no transformation as the tale is radically reworked: Pinocchio does not become a real boy but remains forever a puppet. The closing image is of Pinocchio in a mutual embrace with Geppetto's wife and the Fairy, all of whom are smiling and wide-eyed. In this work, the transformation is denied, yet in a way, the transformation of Pinocchio from puppet to real boy does take place: At the very beginning, as Pinocchio is visually constructed from the outset in humanised form. Herein lies an interesting contradiction between the text and the images, in which the puppet character is visually represented in humanised form – suggesting that he became human as soon as he was crafted – but Pinocchio proclaims in the narrative that he in fact remains a puppet despite being visualised as a boy from the outset. Paradoxically, the ambiguity is sustained in the narrative–illustrative nexus even though neither narrative nor illustrative telling of the tale allows for either ambiguity or transformation independently.

Across these depictions of Pinocchio's transformations, we can identify at least two further contradictions through which Pinocchio is variously constructed. The first concerns the difficulty in sustaining the material and ontological ambiguity of Pinocchio once he has become a real boy, which only the *Folio* version achieves. This raises questions as to the significance of sustaining Pinocchio's ambiguity despite resolving it as commentary on the human condition, reflecting on the various social straightjackets constraining our possibilities for action through Pinocchio's paradoxical immobilisation as a portrait. The second concerns the potentially problematic relation between narrative and visual constructions of Pinocchio, which was most evident in the *Morpurgo* version. This raises questions as to the point of a story that differs so markedly from the original as – whether Pinocchio's dream was ultimately denied or he became human at the point of creation – it is unclear why he should have endured his ordeals. This is not explored in this work: It is simply that the anticipated ending does not happen.

This radical reworking of the tale raises further questions as to the intertextual and intersemiotic translation of the tale whereby the continuance of the tale is reliant upon its deep meaning not significantly changing through its telling in different works and formats (Consolo 2011b). If Pinocchio is accepted as a mythological as well as a literary figure who is defined by his varied metamorphoses (Consolo 2011b), it would seem important that these metamorphoses feature in the retelling of the tale. Yet in several of these works some metamorphoses feature only in the narrative and in one – *Morpurgo* – the single most important transformation in the definition of both the character and the tale of Pinocchio is omitted entirely. Furthermore, these issues speak to questions raised by Collodi as to whether the work can ever be completed (Wilson 2011), as two of those versions that did complete the work as a narrative (by Pinocchio becoming a real boy) failed to sustain the ontological ambiguity at its heart in the process. Arguably, this failure suggests that the tale has not in fact been completed because its orienting device and defining quality has been lost. Only the *Folio* version successfully brought the tale to an end while enabling its definitional ambiguity to live on, by re-objectifying the newly enlivened Pinocchio in portrait form.

## Conceptualising Pinocchio's becoming

The analyses in this chapter have identified multiple contradictions associated with Pinocchio's materiality, sociality and metamorphoses. Those associated with materiality drew out insights into the varied approaches adopted to Pinocchio's subjectivity, establishing a material–subjective continuum bound up with bodily form, while those associated with sociality established both a material–social and a subjective–social continuum. Finally, the analysis of Pinocchio's metamorphoses unearthed two further contradictions: Between narrative and illustrative representations and between resolving and sustaining the ontological ambiguity at the end of the tale. These findings can be shown diagrammatically as in Figure 6.1.

The material–subjective, material–social and subjective–social continua form the central triangle and – in their varied forms – draw on animal, human and puppet associations. The affective impacts run in a band across the centre of the triangle, with the source positioned between material and subjective as the threats are typically posed by individual characters and take material form (e.g., being burnt, cooked, transformed into a donkey), and the target positioned between subjective and social as those in peril might be individuals or groups. The uneven line along the bottom indicates the potential for contradiction between narrative and illustrative representations, which might work positively to sustain Pinocchio's ambiguity, but which can work negatively to inject incongruence into the work. The straight line along the bottom indicates the potential to sustain the ambiguity at the heart of Pinocchio beyond its conclusion through visual mechanisms that extend the ambiguity from Pinocchio to the reader and society.

Putting this to a brief empirical test by examining the episode with the woodpeckers provides reassuring confirmation of the applicability of this schematic. While the visual of this episode in the *Collector's* and *Folio* versions associate Pinocchio only with animals (birds), and the former does not depict the transformation in progress, in the *Folio* version, Pinocchio's nose extends across two pages with 14 birds around it. His gaze is focused on his nose and the birds, and his mouth is slightly open and rounded in amazement. Pinocchio's gaze and expression emphasise the impact that it had upon him, seemingly amazed at his own material form and fascinated to watch it change in front of his eyes. This is a striking image that speaks clearly to Pinocchio's materiality as wooden and also emphasises its uncontrolled animism and his capacity for human cognition and affect. It is striking in its visual, material and narrative significance. The *Morpurgo* and *Daly* versions form an informative contrasting pair as both show human and animal sociality, but their affective tenor differs. In *Morpurgo*, one image shows Pinocchio's mouth open in surprise as he explores his nose with his hands, and another shows the Fairy kneeling behind Pinocchio with eyes closed, head bowed and arms around him in embrace, surrounded by a mass of birds in flight. The single image in the *Daly* version shows Pinocchio with eyes closed and a downcast expression, his hands on his temples and crying, as birds fly around his extended nose. The Fairy's eyes are also closed, but her hand is to her mouth and she is laughing. Both works emphasise human sociality and Pinocchio's subjectivity but as the birds are not shown at work on Pinocchio's nose, they underplay his materiality, placing him firmly on the human side of the puppet/human border. Moreover, the second image in the *Morpurgo* version undermines its affective significance: While the image is affectively impactful, it is somewhat misleading as the Fairy – according to the text – is comforting Pinocchio after his nose has been shortened by the birds, rather than protecting him from imminent attack as suggested – to me at least – by the image. The visual representations of this event, then, illustrate and corroborate several aspects of the foregoing analysis, including ambiguity as to the source of the threat, the diverse corporealities and socialities constructed, and the potential for contradiction between narrative and illustrative representations of the same event.

This emphasis on contradiction is consistent not only with the preceding analyses in this volume but also with the understanding of puppets and puppetness in Collodi's original narrative, yet it develops this understanding further. Collodi ascribes a quality of 'but' to puppets, commenting that 'in the lives of puppets there's always a 'but' that spoils everything' (*Collector's*, p155). However, there is a distinction to be drawn between this 'but-ness' of puppets and their 'puppetness', wherein the former speaks to the proclivity for something to occur to upset an otherwise positive sequence of events (as epitomised by Pinocchio's misadventures) and the latter speaks to the constructive contradictions upon which puppets are constituted. Pinocchio dreams of becoming a real boy and clearly has human capacities for thinking and feeling, but there are also instances when Pinocchio does not feel pain in the same way that humans would, as when he burns his feet while asleep 'as if his feet belonged to someone else' (*Collector's*, p27). Pinocchio is both living and not living. He is mechanical, with a heart that ticks like a clock, yet he is also advised by the ghost of the cricket to do what his heart tells him is right (*Morpurgo*), establishing that his heart is more-than-mechanical. He feels despair at the death of the Fairy (*Collector's*), protective towards his parents (*Morpurgo*) and ashamed of his actions (*Morpurgo*; *Collector's*). Pinocchio, then, is a walking, talking contradiction and a prime example of the grotesque: Wooden but human, living but not alive, mechanical yet sentient. The detailed analyses in this chapter, however, provide greater specificity regarding this constitutive 'but'-ness or puppetness by articulating eight core contradictions on which Pinocchio and his transformations have been variously constructed, presented in Figure 6.1 as a conceptual framework, which is explored in more detail in the next section.

## Spatialising Pinocchio's becoming

While the spatial and temporal indeterminacy within Pinocchio and its emphasis on marginal spaces have previously been addressed (Klopp 2011; Consolo 2011a), my interest here is not in the spaces and places within the tale but in the spatialities generated by the contradictory qualities through which Pinocchio is constructed and his progressive becoming-human is (or is not) established: The micro-spatialities generated by the borderscape. This chapter has identified eight contradictions upon which Pinocchio was variously constructed in the illustrations examined. Four of these related to materiality (flesh/wood, form/expression, body/head and body/mind) and commonly entailed – for example – a wooden form capable of expressing human emotion. Two contradictory continua were associated with Pinocchio's sociality, one of which constructed Pinocchio as part of an animal, human or puppet community and one that operated somewhat indirectly in constructing the puppet, as clarifying or confusing the source and/or target of threat. Two further contradictions were identified that functioned at the level of the work rather than at the level of the puppet protagonist: That associated with maintaining the material and ontological ambiguity that defines the tale, and that between narrative and visual constructions. These, then, can be considered the topological gulfs upon which Pinocchio is constructed in these visual practices of borderscaping, and Table 6.3 summarises both which of these topological gulfs are employed in each work and how consistently Pinocchio was materially and socially constructed through the transformations.

The *Folio* and *Collector's* versions are notable for their comprehensiveness, with both texts covering both transformations (at least in their completion) and the only contradiction not in evidence with respect to Pinocchio's materiality and sociality being absent purely because the *Folio* version does not confine indications of woodenness to the body rather than the face. By contrast, both the *Daly* and *Morpurgo* texts omit one or other of the transformations considered, and both – especially the *Morpurgo* version – are more selective in the contradictions employed to construct Pinocchio. The omission of Pinocchio's animal sociality from *Daly's* telling of the tale, while

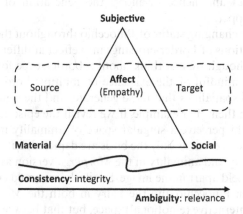

*Figure 6.1* Constructive contradictions of Pinocchio's becoming.
Source: Author.

inconsistent with the original story and constraining the borderscape, does not fundamentally undermine the core ambiguity as to Pinocchio's status as puppet or human in the way that *Morpurgo*'s omission of his puppet sociality does, such as when the other puppets at the puppet theatre do not recognise Pinocchio. This, combined with the visual emphasis on Pinocchio in human-ised form thoroughly de-puppetises Pinocchio, despite his claim to be the most famous puppet in the world. However, this de-puppetisation is in stark contrast to the stated conclusion to the tale, which leaves Pinocchio perma-nently in puppet form, and highlights three inconsistencies: Within the narra-tive itself, between the narrative and visual elements of the work (as explored earlier), and between the original tale and its retelling. More pertinent, here, though, are the implications of this for the borderscape.

As a starting point, this discussion indicates the need to consider the importance of a pre-existing work to the retelling of that work, in other words, what and how big a role the original story should play in the border-scape for a new telling of the same tale. With the *Morpurgo* version, although there is a border evident between human and animal, forging links across that border would constitute the grotesque but not a puppet, as the removal of puppet relationality denies the possibility of this borderscape creating a pup-pet. It cannot be done as the borderscape lacks a key ingredient, making this borderscape self-limiting insofar as the contradictory elements upon which the archetypal transformational puppet could be – and originally was – con-structed are excluded in both narrative and illustrative ways. In contrast to the suggestion that juxtapositions between narrative and illustrative accounts generate an unresolved totality akin to Pinocchio himself (Wilson 2011), there cannot be a totality that constitutes 'puppet' if the materialities and/or socialities established within its generative borderscape preclude it. Pinocchio's proclaimed permanent puppet form is therefore impossible given the borderscape upon which it is nominally predicated, as it denies Pinocchio's puppet attributes and hence disallows the generation of the totality of Pinocchio as puppet.

Moreover, the changing status of Pinocchio throughout the tale brings dif-ferent configurations of border-crossings into effect at different points in the narrative, so although we can think about a borderscape for each work, we also need to be mindful of the existence of multiple bordercapes within a work, the micro-spatialities that these generate, and the significance of con-sistency between them. For example, if we revisit the episode with the wood-peckers, we might perceive a singular space of animality in the *Collector's* visualisation of this event as only the birds are depicted, but a binary, parallel space of humanity and animality in the *Morpurgo* version as the two sources of sociality are held apart in the image. Similarly, we could consider the rela-tionship between Pinocchio and the Fairy in both the *Morpurgo* and *Daly* versions as an interactive (emotional) space, but that between Pinocchio and the birds in the *Folio* version as a reflective space. Significantly, the illustrative detail in the latter transforms this space from one of mere interaction, as Pinocchio's observation of the activity of the birds on his nose rebounds onto him in reflecting upon his own being, which unites his material reality, his bodily aspirations, his animal associations and his unique subjectivity. These conceptualisations can be associated with different aspects of the constructive contradictions (see Figure 6.1), with the singular and parallel spaces arising from varied material and social associations, the interactive space as subjective, social and affective, and the reflective space as returning to material and social associations but with an affective flavour. These different ways in which aspects of those contradictions are operationalised (how puppetness is interrogated), then, generate different (material and social) spatialities, with implications for Pinocchio's subjectivity and status. As those spatialities are grounded in the constructive contradictions that characterise the borderscape (in different combinations and with different emphases), we can approach the same issue from the perspective of either the spatialities that emerge or the borderscape/s from which they emerge, but human–puppet spatialities can only emerge from a borderscape that allows for them.

As the pertinent constructive contradictions change even within a single work as the story progresses, the spatialisation emergent from this is also fluid in nature. As Pinocchio's adventures and becoming unfold, the micro-spatialities generated by the individual constructions of each event change, and these might be entirely discrete or might overlap with each other. Consistency becomes important here, as micro-spatialities that overlap due to sustained material associations might conflict if their social or subjective associations differ, and vice versa, as with *Morpurgo's* Pinocchio claiming to be a famous puppet yet being visually represented – along with the other puppets – in very human form and not being recognised by the other puppets. In this instance, the humanised puppets seem out of place in the puppet theatre, Pinocchio self-identifies as a puppet yet seemingly is not one and his animal-social status is equally conflicted, undermining his ontological ambiguity despite not resolving it, because the possibility of a totality that incorporates 'puppet' is constitutively precluded. We can thus distinguish between the narrativised places in which the story unfolds, the borderscapes through which Pinocchio is constructed, and the implicit micro-spatialities generated by individual events within the tale, yet they all intersect in the construction of Pinocchio, making consistency between them important to the effectiveness of both the tale being told and the character being constructed.

As such, micro-spatialities within a single work are nested within the illustrative ontological space of Pinocchio's being (the totality of visual constructions of Pinocchio throughout that volume); we have a similar scalar relationship between the micro-spatialities and a macro-spatiality as we have between the different levels of borderscape. Consistency again comes to the fore, here, as conflict at the level of micro-spatialities could impact the integrity of the overall construction of Pinocchio, and inconsistency in the underpinning borderscape is likely to generate inconsistency between the micro-spatialities that emerge from it. Moreover, the relationship between the narrative and illustrative content within a single work introduces a potential for further contradictions between the micro-spatialities generated in each version of the tale, perhaps reinforcing the ambiguity at the heart of the tale but also risking pushing this to the point of incongruence that undermines the story. Further, the intertextual and intersemiotic translation of the Pinocchio tale into successive retellings complicates this spatialisation and means that Pinocchio continues his process of becoming between these proliferating works, bringing us back to questions as to the role that an original work could or should play in the borderscape of a new work, with implications for each reconstruction of Pinocchio in sustaining, denying or undermining Pinocchio's ambiguity.

Evidently, Pinocchio undergoes myriad micro-transformations within each work and macro-transformations between works, both ontological and spatial, in an ongoing process of becoming, even if he does not ultimately transform into a real boy. These ontological and spatial transformations are mutually constitutive. He is constructed on the contrariness at the heart of puppetness, building upon Collodi's 'but' of the puppet and identified in this visual analysis as involving eight sets of contradictions, and the spatialities generated by this puppetness can reciprocally help to substantiate, sustain or undermine it. Pinocchio is indeed the archetypal puppet: Constructed on contradictions and perpetually becoming through ontological and associated spatial transformations.

## Conclusion

Despite telling essentially – and in some cases precisely – the same story the visual depictions varied significantly across the works analysed. Examining these constructions in relation to Pinocchio's materiality, sociality and metamorphoses revealed a series of contradictions that are variously employed to construct specific depictions of Pinocchio. Interrogating Pinocchio's materiality revealed contradictions of flesh/wood, form/expression, body/head, body/mind that are differently configured across the four works. The study of his sociality brought to light associations not only with human and puppet communities but also animals and highlighted how confusion regarding these associations can detrimentally affect the affective impact of the image, formalising these as social and affective contradictions. The analysis of his metamorphoses revealed two further contradictions: That between the narrative and illustrative constructions in a work, and that between resolving Pinocchio's ambiguous status and sustaining that ambiguity at the end of the tale. The first of these was found to hold significance for the integrity and power of the work, whereby inconsistency between narrative and illustrative constructions might enhance the sense of ontological ambiguity but might also confuse or complicate it excessively. The second was found to hold significance for the relevance of the Pinocchio story to contemporary society, as sustaining the ambiguity even after Pinocchio has become (or not) a real boy strengthens the relatability of the story, its meaning and its morality to the lived circumstances of the reader.

These contradictions were developed into a conceptual framework oriented around three continua between social, material and subjective factors and articulated as a suite of topological gulfs across which Pinocchio was variously constructed. Subsequently, the effectiveness with which these constructive contradictions were employed by each of the four works was examined, which distinguished the *Morpurgo* version as diverging significantly from the others in its visual emphasis on Pinocchio in human form and in its disavowal of Pinocchio's puppet sociality, which was deemed to have undermined the borderscape and precluded the construction of Pinocchio as a puppet, which is especially paradoxical given that the conclusion leaves Pinocchio permanently a puppet. This also led to a discussion of the nesting of more specific borderscapes within the borderscape for the work as a whole, as different combinations of constructive contradictions come to the fore at different points in the narrative, and consideration of the significance of in/consistency between them.

Considering these constructions in spatial terms brought to light a different perspective on how Pinocchio is constructed. As with the borderscapes, inconsistency between the spatialities that emerge from them can detrimentally affect both the construction of Pinocchio and the integrity of the tale due to the nesting of these micro-spatialities of individual relational events within the overall ontological space of the contradictory character, both within and between works. As the archetypal puppet, Pinocchio's constructedness provides insight into the very notion of puppetness, and the foundational nature of the contradictions identified was related to Collodi's own articulation of puppetness as the 'but' of the puppet. In providing more detail about specific constitutive forms that this 'but' – reformulated as puppetness – can take, this chapter both develops our understanding of puppetness and schematises these contradictory constructions in Figure 6.1. It establishes that Pinocchio is not only the archetypal puppet in contrariness and transformative capability – even if he does not ultimately transform – but also that he is spatially constructed through myriad micro- and macro-spatialities of puppet contradictions in a perpetual process of becoming.

## References

Consolo, S. (2011a) The watchful mirror: Pinocchio's adventures recreated by Roberto Benigni. In: K. Pizzi (ed) *Pinocchio, puppets and modernity: The mechanical body.* Routledge: New York, 175–199.

Consolo, S. (2011b) The myth of Pinocchio: Metamorphosis of Collodi's puppet from pages to screen. In: K. Pizzi (ed) *Pinocchio, puppets and modernity: The mechanical body.* Routledge: New York, 163–174.

Klopp, C. (2011) Workshops of creation, filthy and not: Collodi's Pinocchio and Shelley's Frankenstein. In: K. Pizzi (ed) *Pinocchio, puppets and modernity: The mechanical body.* Routledge: New York, 63–73.

Lucas, A.L. (2011) Puppets on a string: The unnatural history of human reproduction. In: K. Pizzi (ed) *Pinocchio, puppets and modernity: The mechanical body*. Routledge: New York, 49–61.

Perrot, J. (2011) Carlo Collodi and the rhythmical body: Between Giuseppe Mazzini and George Sand. In: K. Pizzi (ed) *Pinocchio, puppets and modernity: The mechanical body*. Routledge: New York, 17–47.

Pizzi, K. (2011) Introduction. In: K. Pizzi (ed) *Pinocchio, puppets and modernity: The mechanical body*. Routledge: New York, 1–15.

Riva, M. (2011) Beyond the mechanical body. In: K. Pizzi (ed) *Pinocchio, puppets and modernity: The mechanical body*. Routledge: New York, 201–214.

Wilson, S. (2011) Unpainting Collodi's fireplace. In: K. Pizzi (ed) *Pinocchio, puppets and modernity: The mechanical body*. Routledge: New York, 109–133.

# 7 Humans becoming puppet

In this final substantive chapter, my attention turns to the opposite transformation, that is of humans becoming puppets. As the analysis of the musical works in Chapter 4 incorporated consideration of such transformations to identify the ways in which puppets were constructed, my focus here returns to the literary and filmic works examined in Chapters 1–3. Reviewing these works, ten directly and substantively address such a transformation, to which I have added four works that I have sourced more recently (asterisked below) to generate the following list of works analysed:

Films:

- *Dead Silence* (2007) dir: J. Wan.
- *Robert* (2015) dir: A. Jones.
- *Dolls* (1986/2014) dir: S. Gordon.
- *Being John Malkovich* (2003) dir: S. Jonze.
- *The Happytime Murders* (2018) dir: B. Henson.
- *The Beaver* (2011) dir: J. Foster.
- *Dead of Night* (1945) dir: A. Cavalcanti et al.
- *Judy and Punch* (2019) dir: M. Foulkes.
- *Puppet Master 3: Toulon's Revenge* (1991) dir: D. Decoteau.
- *Puppet Master: The Littlest Reich* (2018) dir: T. Wiklund and S. Laguna.

Books:

- Aaronovitch, B. (2011) *Rivers of London*. Gollancz: London.
- Wynn Jones, D. (2008) *The Magicians of Caprona*. HarperCollins: London.
- Carter, J. (2016) *The Death of Mr Punch*. Peter Owen: London and Chicago.
- *Tangredi, A. (2012) *The Puppet Master's Bones*. Viverridae Press: Los Angeles, CA.

The chapter is structured in three sections. In the first section, I present a comparative analysis of these works in relation to the understandings of

DOI: 10.4324/9781003214861-16

puppets that have emerged through the preceding chapters, through which different perspectives are identified on aspects of the dimensional framework, constructive contradictions and ideas of borderscaping. The implications of this analysis are then explored in terms of a schematic of different perspectives that the works adopt towards the transformation of humans into puppets, through which a set of constructive complementarities are elaborated that supplement the constructive contradictions already identified. Thirdly and finally, I dig deeper into these perspectives to unearth ten specific devices that are employed to transform humans into puppets and use these to conceptualise a suite of corporealities that work either in isolation or in combination to configure these diverse human-to-puppet transformations.

Subsequently, these are re-articulated as a series of ecologies and technologies – oriented towards human and puppet characteristics or associations respectively – and I use these to integrate the varied conceptualisations that have emerged through the analyses and to elaborate more fully on the definition of borderscape and the practice of borderscaping. Thus, this chapter generates a thorough and detailed – though not exhaustive – account of the ways in which human-to-puppet becomings are constructed, draws together multiple analytical threads from earlier chapters and specifies the borderscape. These will then be considered in tandem with the constructions of Pinocchio's puppet-to-human becomings from the previous chapter in the Conclusion to Part 3, to establish a composite schematic of the varied ecologies and technologies of puppet-becomings, where puppet-becomings encapsulates transformations in both directions.

## Comparative analysis

I begin by analysing these works involving the human-to-puppet transformation in relation to the dimensional framework and constructive contradictions previously identified, focusing on materiality and sociality as the two most prevalent categories. Table 7.1 presents a summary of this analysis. This establishes that despite utilitarian, relational and qualitative dimensions being readily applicable to each work and different combinations of materiality and sociality being identifiable, there are also many and varied new elements in how these works construct the human-to-puppet transformation as distinct from how they construct puppets. While some of these are directly attributable to the genre of the work, as explored in Chapter 2, others suggest that there are several means by which the human–puppet transformation is established that have not been fully unearthed in the preceding analyses. These can be clustered under three headings: (1) the morphing of bodily capacities, including movement and voice, which emerged in Chapter 3 but has not yet received further attention; (2) the significance of puppet plays, in the context of which the transposition of name or identity becomes a key signifier of transformation, which builds on the significance of intertextuality identified in relation to Pinocchio in Chapter 6; and (3) self-actualisation, which goes beyond the puppet being a social mediator or a tool for

*Table 7.1* Comparative analysis of human-to-puppet transformations

| Work | Outline of work | Dimensions | Contradictions | New elements |
|---|---|---|---|---|
| *Dead Silence* Film (2007) dir: J. Wan | A vengeance tale in which a puppet maker invests her own spirit in her puppets and seeks retribution by turning her victims into puppets. | Utilitarian: narrative role focal and full, indirect moral impact<br>Relational:<br>• Social, constitutive<br>• Performative impact: focal<br>• Borderscaping varied<br>Qualitative: moderate puppetness, creative ontology and grotesquery | Materiality:<br>• Flesh/wood<br>Sociality:<br>• Human<br>• Puppet | • Spirit/body<br>• Transformation from corpse to puppet<br>• Appears human but with the mobility of a puppet<br>• Puppets as family |
| *Robert* Film (2015) dir: A. Jones | A vengeance tale in which the spirit of a murdered boy is invested in a puppet, which seeks violent retribution, with the spirit then transferring to the puppet's new child companion. | Utilitarian: narrative role focal and full<br>Relational:<br>• Individual, informational<br>• Borderscape deleted<br>Qualitative: moderate puppetness, creative ontology, creative grotesquery (character contorted) | Materiality:<br>• Material/ form<br>Sociality:<br>• Human | • Spirit/body<br>• Innocence/malevolence<br>• Chain of human–puppet–human possession |
| *Dolls* Film (1986/2014) dir: S. Gordon | A dark comedy in which an elderly couple nurture the love of toys in adults by turning into toys (including puppets) any adults failing to show sufficient reverence for toys. | Utilitarian: narrative role focal and full, indirect moral impact<br>Relational: social, constitutive<br>Qualitative: moderate puppetness, creative ontology and grotesquery | Materiality:<br>• Body/head<br>• Flesh/wood<br>Sociality:<br>• Human<br>• Puppet (toy) | • Innocence/malevolence<br>• Play/punishment<br>• Puppet affect (fear) |

*(Continued)*

*Table 7.1* (Continued)

| Work | Outline of work | Dimensions | Contradictions | New elements |
|---|---|---|---|---|
| *The Magicians of Caprona* Book (2008) auth: D. Wynn Jones | A magical tale in which an evil Duchess turns members of rival households into puppets from the Punch and Judy show, forcing them to act out the play's violence against each other. | Utilitarian: narrative role focal and full Relational: social, constitutive Qualitative: moderate puppetness, creative ontology | Materiality: <br>• – <br>Sociality: <br>• Human | • Size, not form/material <br>• Transposition of name/ identity <br>• Control through implement, not body <br>• Spirit/body <br>• Puppet affect (pain) <br>• Context of puppet play |
| *Being John Malkovich* Film (2003) dir: S. Jonze | A fantasy tale in which a puppeteer finds a portal from his office into John Malkovich's brain, leading to the occupation of John Malkovich by other people who control him as a puppet. | Utilitarian: narrative role focal and full Relational: <br>• Individual, psychological <br>• Social, constitutive <br>• Borderscaping varied <br>Qualitative: moderate puppetness, transformational ontology, creative grotesquery (character contorted) | Materiality: <br>• Body/mind <br>Sociality: <br>• Human | • Vocal control <br>• Being in another's skin (not the puppet but who it represents) <br>• Control from within <br>• Bodily congruence <br>• Self-realisation |
| *The Happytime Murders* Film (2018) dir: B. Henson | An adult Muppet murder mystery in which puppet and human communities live side by side but with their own social practices and codes, and in which a human police officer has received a 'fluffy' puppet liver transplant. | Utilitarian: narrative role indirect/ focal and full, indirect moral impact Relational: <br>• Individual, informational <br>• Social, discrete <br>• Borderscape deleted <br>Qualitative: moderate puppetness, conventional ontology, creative grotesquery | Materiality: <br>• Form/ material <br>Sociality: <br>• Human <br>• Puppet | • Surgical bodily merging <br>• Attitudinal change |

| | | | | |
|---|---|---|---|---|
| *Rivers of London* Book (2011) Auth: B. Aaronovitch | A supernatural vengeance tale in which the spirit of Mr Punch sequesters the bodies of individuals in the present day to act out his own play. | Utilitarian: narrative role indirect and partial (Punch as person not puppet)<br>Relational:<br>• Social, constitutive<br>• Borderscape deleted (or overturned)<br>Qualitative: negligible puppetness, creative ontology, transformational grotesquery | Materiality:<br>• Material/form<br>• Head/body<br>Sociality:<br>• Human | • Spirit/body<br>• Transposability of name/identity<br>• Context of puppet play |
| *The Beaver* Film (2011) dir: J. Foster | A psychological tale about a successful executive who falls into depression and attaches to a hand puppet of a beaver, in which he progressively invests himself but through which he then finds himself. | Utilitarian:<br>• Narrative role focal and full<br>• Focal performative impact<br>Relational:<br>• Individual, psychological<br>• Borderscaping varied<br>Qualitative: moderate puppetness, conventional ontology, Transformational grotesquery | Materiality:<br>• Body/mind<br>Sociality:<br>• Human | • Bodily extension<br>• Transposition of identity/name<br>• Salvation<br>• Self-realisation |

*(Continued)*

*Table 7.1* (Continued)

| Work | Outline of work | Dimensions | Contradictions | New elements |
|---|---|---|---|---|
| *The Puppet Maker's Bones* Book (2012) Auth: A. Tangredi | A mystery tale about a human with a rare condition that makes his physical touch fatal to others, prompting the confinement of his familial and social relations to the puppet world and leading to him immortalising his lost loved ones in puppet form. | Utilitarian:<br>• Narrative role focal and full<br>• Performative focal impact<br><br>Relational:<br>• Individual: psychological (affective facilitator)<br>• Social, constitutive<br>• Borderscaping varied and complex<br><br>Qualitative: substantive puppetness, creative ontology | Materiality:<br>• Material/form<br>• Body/mind<br><br>Sociality:<br>• Human<br>• The afflicted<br>• Puppet | • Behaviour and functionality<br>• Human interaction through puppets<br>• Puppets as family |
| *The Death of Mr Punch* Book (2016) auth: J. Carter | A tragi-comic tale in which a former Punch and Judy Professor struggles to distinguish his identity between Mr Punch the puppet that he used to operate and himself as the puppeteer, as he battles with dementia. | Utilitarian: narrative role full but incidental (Punch as person not puppet)<br>Relational:<br>• Individual, psychological<br>• Social, constitutive<br><br>Qualitative: negligible puppetness, conventional ontology, creative grotesquery (character contorted) | Materiality:<br>• Body/mind<br><br>Sociality:<br>• Human | • Transposition of name/identity<br>• (Lack of) self-realisation |

| | | | |
|---|---|---|---|
| *Dead of Night* Film (1945) dir: A. Cavalcanti, C. Crichton, B. Dearden and R. Hamer | A suspense tale. A man hears uncanny tales from his hosts at a retreat, including one involving the alleged murder of one ventriloquist by another, but the dummy/puppet committed the crime. The alleged perpetrator merges with the dummy. | Utilitarian: narrative role incidental and partial Relational: individual, informational Qualitative: moderate puppetness, conventional ontology | Materiality: • Body/mind Sociality: • Human | • Merging of voice |
| *Judy and Punch* Film (2019) dir: M. Foulkes | A dark comedy in which Punch and Judy (as people) are puppeteers of a Punch and Judy puppet show. Their lives mirror the Punch and Judy story but Judy is not killed and wreaks her revenge on Punch. | Utilitarian: narrative role incidental but full. Indirect moral impact Relational: social, constitutive Qualitative: negligible puppetness, conventional ontology | Materiality: • Body/mind • Material/ form Sociality: • Human | • Transposition of names and identities • Context of puppet play • Chain of puppet– human–puppet influence |
| *Puppet Master III: Toulon's Revenge* Film (1991) dir: D. Decoteau | In 1940s' Germany, the Reich seeks to return the dead into living form for military purposes by stealing Toulon's serum that turns people into puppets, enabling them to live forever as a puppet. Toulon seeks revenge. | Utilitarian: narrative full and focal Relational: constitutive social, con-joined borderscape Qualitative: moderate puppetness, creative ontology and grotesquery | Materiality: • Material/ form • Flesh/wood Sociality: • Human • Puppet • The deceased | • Militarisation • Historic context • Medical serum • Puppets as family |

*(Continued)*

*Table 7.1* (Continued)

| Work | Outline of work | Dimensions | Contradictions | New elements |
|------|-----------------|------------|----------------|--------------|
| *Puppet Master: The Littlest Reich* Film (2018) dir: T. Wiklund and S. Laguna | Prior to an auction of Toulon's puppets, they come to life and launch a vicious killing spree targeting Jewish and homosexual characters. The protagonist must find a way to deactivate them. | Utilitarian: narrative use full and focal Relational: individual informational, social discrete (mostly) and borderscape (mostly) deleted Qualitative: moderate puppetness, some creative ontology and grotesquery | Materiality:<br>• Flesh/wood<br>• Material/ form<br>Sociality:<br>• Human<br>• Puppet | • Puppets versus dolls<br>• Militarisation<br>• Puppet joins unborn baby in the womb; puppetised baby Führer<br>• Historic context |

communication to the human becoming more human through the process of becoming puppet.

While the emphasis in previous analyses on bodily form and materiality is equally evident in the human-to-puppet transformations depicted in these works, there is also a greater prevalence of the morphing and merging of bodily capacities, most notably in relation to movement and voice. Within horror films, the reactivation of a corpse by turning it into a puppet is both complementary and oppositional to more common approaches of spirit possession, magic or curse. It is complementary to such approaches in the sense that in each case, the puppetised character is often rendered incapable of voluntary movement, but it is oppositional in the sense that the puppetised form of the human remains capable of humanistic movement. However, this is established in different ways. In *Dead Silence*, the revelation of the human character as having been puppetised long before the point of revelation establishes the lifelike nature of their puppetised movement and mannerisms, thereby masking any recognition of the puppet–human distinction until the point of revelation in a radical dissolution of the border between puppet and human. In *The Puppet Maker's Bones*, the utilisation of the bones of the puppet maker's family members in the material construction of his puppets establishes the capacity for this human–puppet to move and behave in entirely humanistic ways, and positions those puppetised family members as being manipulated in precisely the same way in which they used to manipulate their own puppets while still alive. By employing the materiality of the human to facilitate the smooth and realistic movement of its puppetised form, the differences in both materiality and movement between puppet and human are elided as it becomes impossible to distinguish one from the other, establishing a new constructive contradiction between movement and body. In these examples, humans become puppets yet retain their apparent and material humanity, and in clear contrast to the everyday understandings of puppets that clashed with those of puppet practitioners (see Chapter 1), the naturalness and fluidity of movement maintained with these puppets are much more consistent with the expectations and perspectives of practitioners.

Similarly, although bodily merging through material recombination, surgical procedure and facial transfiguration are consistent with previous analyses, three of the works analysed here draw attention to the voice as a defining characteristic of the puppet that the human is destined to become. While in *Being John Malkovich*, this involves the sense of controlling the words that one utters rather than the tonality of the voice, in both *Dead of Night* and *Rivers of London*, the vocal pitch and mannerisms transfer from puppet to human as the human becomes more puppet. In *Rivers of London*, not only does the sequestered human character articulate Mr Punch's famous line 'That's the way to do it!' but also takes on his high-pitched squawking elocution intermittently with the more bass tones of Henry Pyke, Mr Punch's alias. In this instance, again, we need to think beyond contradictions of form and materiality, extending our interest to the borders and contradictions between bodily *capacities*.

Related to the significance of bodily capacities is the facilitation and enhancement of psychological or cognitive capacities, which finds expression in the form of human self-actualisation. Taking us beyond the role of the puppet in facilitating communication to enabling self-realisation or salvation, the human characters become – in some way – a 'truer' form of their former self. *Being John Malkovich* serves as an example here as the experience of being the puppeteer inside John Malkovich's head is described in revelatory terms in enabling the occupant not only to see the world in a new light through John Malkovich's eyes but also to see themselves in a new way; to understand who they are either differently or more fully than had previously been the case. While it could be argued that this is more a case of the puppet master – rather than the puppet – securing personal growth as the puppet is John Malkovich himself, the puppet master is inside John Malkovich's head so is simultaneously both the puppet and its master and is seeing the world through the puppet and with its eyes. In *The Beaver*, Tom Hanks' human protagonist progressively invests himself in the glove puppet of a beaver that he took to wearing after succumbing to severe depression. Refusing to remove the beaver puppet even while showering or during sexual intercourse, the human character takes on the name of the titular animal and is rendered incapable of communicating when forced to remove the puppet, while the puppet takes on the role and duties of the human protagonist in his professional life and excels in that role. He first becomes a puppet, then becomes a phenomenon in his professional life as that puppet and only then goes on to become his true self as an individual, a husband and a father, once liberated from the puppet. While remaining no more than an accessory to his human sidekick in bodily terms, the cognitive and psychological capacities of the human become progressively enabled by and facilitated through the material extension of the beaver glove puppet as the human becomes puppet becomes human. In neither case can the puppet be said to mediate the social acceptability of the experiences articulated and while in *The Beaver* the glove puppet does facilitate communication in a psychological sense (consistent with the analysis presented in Chapter 3), this is at the expense of a previous ability to communicate without the aid of the puppet, so it constitutes a replacement rather than an enhancement of communicative ability (thereby supplementing the analysis in Chapter 3). There is, though, a reconstitution of self that comes about as each human character integrates progressively with the puppet that they become, so becoming puppet – despite conventional tropes suggesting otherwise – can enable becoming (more) human in a conjoined or reciprocal becoming, establishing such works as classic grotesquery in prompting critical consideration of what it is to be human (Connelly 2012; Pilný 2016).

Finally, the most common feature of these works that stands out from the previous analyses is the role of key puppet plays and characters in establishing the context for the transformation that takes place, reintroducing questions from Chapter 6 concerning the degree to which a new telling should diverge from the parameters set by the original story. In all four examples that evidence this contextualisation it is the anarchic Mr Punch and his vio-lent rebelliousness that provides such context. *The Death of Mr Punch* frames the cognitive decline of a former Punch and Judy Professor in the context of the Punch and Judy story, complete with colourful costume, shrill voice, slap-stick and bell in a psychological process of becoming puppet that impacts appearance, voice and behaviour. In *Magicians of Caprona*, representatives of two rival households are turned into Punch and Judy puppets by magic and are forced into acting out the play and battling with each other by a mali-cious Duchess. While these characters do not self-identify with Punch or Judy, their appearance and behaviour are again transformed into those of their respective puppetised characters. However, it is *Rivers of London* and *Judy and Punch* that are most interesting in this context as they each inter-sperse the historical context of Mr Punch with the fictional narrative of the respective tale, and they each entail the transposition of names and identities between the 'real' Punch, the Punch puppet and the puppetised human, but they do so in different ways.

In *Rivers of London*, the spirit of Henry Pyke (aka Mr Punch) from history sequesters the bodies of humans in the present to enact his violent revenge on unsuspecting innocents, transfiguring their faces into the spitting image of the infamous puppet in the process, thereby transforming the human targets into puppets both visually and behaviourally. Most telling about this work, though, is the plan devised to put an end to his misdemeanours. The protago-nists plan to put themselves (not as puppets) into the Punch and Judy story to track down the bones of Henry Pyke (thus as a human, not a puppet) and destroy them. Here, not only does the context of the Punch and Judy story provide the parameters for the visual, behavioural and vocal human-to-pup-pet transformations that take place, but it provides the means by which the enforced re-enacting of the story can be ended even while those participating in the puppet play are doing so as people and not puppets. Yet, by diligently following the Punch and Judy plot, they are – in effect – puppetised as their behaviour is being manipulated by the play. This novel, then, involves both bodily reconfiguration (as explored in Chapter 1) and control through con-text (as discussed in Chapter 2) and bidirectional transformations in an intri-cate contradictory intersplicing of puppet and human. People in the present become puppets, police work becomes a puppet play that is not played by puppets and the ghost of the puppet (Pyke/Punch) is put to rest by destroying the materiality of his non-puppet (formerly living) form.

In *Judy and Punch*, the two titular human characters run a marionette pup-pet show of Punch and Judy and inadvertently live out the Punch and Judy story in their own lives, but with a twist whereby Judy is not quite killed by Punch and later manages to exact revenge by chopping off his hands. In this complex interweaving of the original puppet story and its retelling in the 'here and now' of the narrative, the parallels drawn between the violence in the puppet play and that exhibited by the society in which it is being per-formed are reminiscent of the social relevance of puppet narratives uncov-ered in Chapters 2 and 3. At the same time, the unconventional ending reinserts the Punch and Judy story into the real history of Punch's cultural evolution. This tale provides an (albeit fictional) account of the puppet Punch's own transformation from a marionette to a glove puppet through the human Punch's sudden inability to operate marionettes in the absence of his hands once Judy has exacted her revenge, requiring the puppet's evolution into glove puppet form, which is its more familiar form in contemporary Western society. Here, the context of the play, the history of the play and the narrative of the story are intricately interwoven in a series of constructive complementarities rather than constructive contradictions. Beyond the bodily, material and behavioural transformation from humans to puppets, then, the context of this puppet play provides its own varied means of transforming humans into puppets.

Collectively, these alternative constructions of puppets and human-to-puppet transformations generate an additional set of constructive contradictions: (1) that between movement and body (in either materiality or form) and (2) that between bodily capacities as much as bodily forms and materialities. In addition, this initial analysis of human-to-puppet becomings highlights the need to think about both the potential for such a process of becoming puppet to have beneficial as well as detrimental implications for the puppetised person if they simultaneously become more human, and the constitutive role played by specific puppet characters and plays in establishing the conditions of possibility for the puppet that the human can become. These latter two can both be considered constructive complementarities, the first of which relates to the person or character and the second to the play or plot, providing an alternative perspective on both the construction of puppets and the grotesque as being constituted not only in contradictions but also in complementarities. However, this is not to suggest that constructive contradictions and constructive complementarities are mutually exclusive, as the complementarity provided by a pre-existing puppet plot might be enacted by establishing the new plot in contradiction to the original (as was the case in Morpurgo's version of Pinocchio in Chapter 6). This might or might not be enacted successfully, but the dual construction of puppets through diverse sets of contradictions and complementarities provides for a fuller understanding of such constructions.

## Transformational progressions

Also evident across the works analysed was a series of topical categories (e.g., perspective, behaviour, theme), each of which was associated with a pair of alternative approaches to the human-to-puppet transformation and each of which is associated with its own spatiality. These are summarised in Figure 7.1, showing how these paired approaches progress through the topical categories, as the human (to the left) becomes a puppet (to the right).

Within the category of perspective, the human concerned exhibited either a positive (or increasingly positive) attitude towards puppets or an affinity with puppets, generating a spatiality of directedness towards the puppet/s.

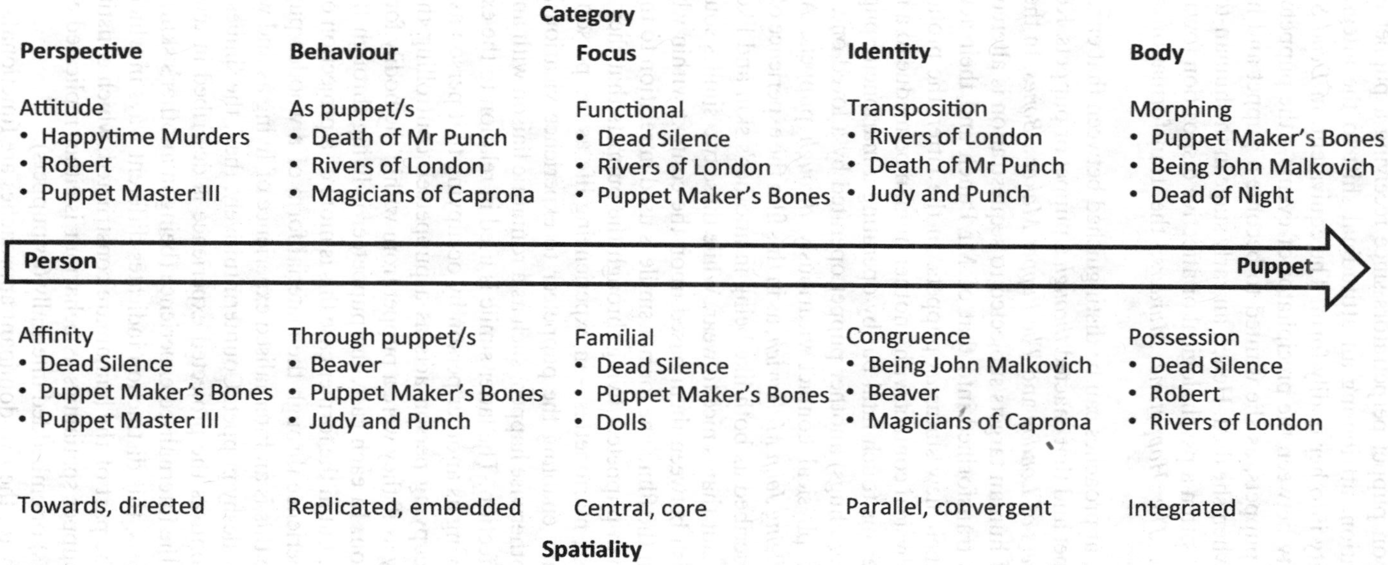

| Perspective | Behaviour | Category | | | Spatiality |
|---|---|---|---|---|---|
| | | Focus | Identity | Body | |
| **Attitude**<br>• Happytime Murders<br>• Robert<br>• Puppet Master III | **As puppet/s**<br>• Death of Mr Punch<br>• Rivers of London<br>• Magicians of Caprona | **Functional**<br>• Dead Silence<br>• Rivers of London<br>• Puppet Maker's Bones | **Transposition**<br>• Rivers of London<br>• Death of Mr Punch<br>• Judy and Punch | **Morphing**<br>• Puppet Maker's Bones<br>• Being John Malkovich<br>• Dead of Night | |
| **Affinity**<br>• Dead Silence<br>• Puppet Maker's Bones<br>• Puppet Master III | **Through puppet/s**<br>• Beaver<br>• Puppet Maker's Bones<br>• Judy and Punch | **Familial**<br>• Dead Silence<br>• Puppet Maker's Bones<br>• Dolls | **Congruence**<br>• Being John Malkovich<br>• Beaver<br>• Magicians of Caprona | **Possession**<br>• Dead Silence<br>• Robert<br>• Rivers of London | |
| Towards, directed | Replicated, embedded | Central, core | Parallel, convergent | Integrated | |

Person ⟶ Puppet

*Figure 7.1* Approaches to human-to-puppet transformation.

Source: Author.

For example, the human police officer in *The Happytime Murders* is said to have 'gone soft' on puppet perpetrators since receiving a puppet liver in a transplant operation, attributing an attitudinal shift to the integration of puppet materiality into her bodily form. The equivalent, in *Dead Silence*, is an evident affinity between the precipitator of events – the puppeteer Mary Shaw – and her puppets, as she wanted to become a puppet and had them buried with her when she died. Hence, an early stage in the human-to-puppet transformation is often a psychological or affective disposition towards puppets, even if in *The Happytime Murders*, the transformation was not completed.

Behaviourally, approaches can be distinguished between that enacted by a human *as* a puppet and that enacted *through* a puppet or puppets. Key examples here are *Rivers of London* and *The Puppet Maker's Bones*. In the former, the behaviour of human targets subjected to sequestration is altered as their physical form is transformed into that of Mr Punch and their actions are determined by their new status *as* puppets. In the latter, the protagonist is unable to have physical contact with another human being due to a rare condition that makes his touch fatal but by operating a marionette puppet that interacts with (e.g., hugs) another puppet operated by a loved one, he gains the intimacy of physical contact vicariously *through* puppets. A further example here is *Being John Malkovich*, as in this film the experience of becoming puppet is described as both like being in another's skin and like putting on an expensive suit that is nice to wear. While these two similes sound alike, we can distinguish between them based upon the context within which they are articulated in the film. The former simile is used in relation to the experience of the human puppeteer living through the puppets that they manipulate – in this case, marionettes – in experiencing life as the person that the puppet represents, enabling the puppeteer to experience vicariously events that would not otherwise happen, such as a romantic liaison with an unavailable target of affection. The latter simile is used in relation to the experience of one person living as another person by occupying that person as a vessel, whereby the occupying person acts as a puppeteer, controlling the person that they occupy as if they were a puppet from within the bodily form of the puppetised person. In each case, the puppeteer is living through the bodily form of another, but in the former case this is more of a projection of desired or imagined experience through the material form of a wooden puppet and in the latter case this is an internalised experience of living as and within the bodily form of a fleshy puppet. Counterintuitively, then, the similes used are a bit back-to-front, as the projected experience is described in terms more appropriate for the internalised experience (being in another's skin) and vice versa. Nonetheless, this distinction indicates different ways in which puppet behaviours on the part of the human are constructed, which constitutes different human–puppet spatialities as behaviour is either replicated externally (*through* puppets) or embedded internally (*as* puppets).

In terms of focus, the two dominant approaches are functional – with the emphasis on what a puppet can do in comparison to a human – and familial

– with the emphasis on the relations between puppets (and humans) – which generate a focal spatiality in which the narrative revolves around the chosen theme, although these two often go together. For example, in *Dead Silence*, the functionality of puppets lies in their ability to act as repositories of human souls or spirits, which vitalises them to exact the murderous revenge of their former puppeteer on those she holds responsible for her own downfall. At the same time, the puppets are characterised as proxies for the puppeteer's children that she does not have: 'beware Mary Shaw, who had no children, only puppets'. Similarly, in *Puppet Master III: Toulon's Revenge*, the puppets are framed as the family of the puppet master (Toulon) and are the puppetised forms of now-deceased friends and family, immortalised as puppets using a special serum. In *The Puppet Maker's Bones*, the protagonist first fulfils his family life by interacting with his parents through his puppets in lieu of interpersonal bodily contact and then uses the bones of his dead relatives to make his own lifelike puppets, thereby enabling him to have physical contact with his parents but only after their deaths and only as puppets, providing the clearest example of conjoined functional and familial focus.

Under the category of identity, the human-to-puppet transformation is framed as either congruence or transposition. With congruence, the identity of the person and the identity of the puppet come together, as in *The Beaver* when the human protagonist takes on the identity of the beaver puppet while continuing in his own professional role and context. With transposition, the name and behaviour of the puppet might be transferred to the human but the human does not lose their own identity, as in *Rivers of London* when the human police officer is sequestered by the spirit of Mr Punch, which overrides her own character. However, *The Death of Mr Punch* is an interesting counterpoint here, as the impacts of dementia make it difficult to ascertain whether the identity of Mr Punch as the puppet replaces the identity of the former Punch and Judy Professor or reinforces it. The spatiality of this category, then, is either one of parallelism between the human and puppet characters, or of convergence between them. Finally, within the category of the body, and as already outlined, the two dominant approaches are either a morphing of puppet and human bodies and/or bodily capacities or the possession of a human body by a puppet or puppetised spirit, which is most clearly evidenced in *Puppet Master: The Littlest Reich*, in which a puppet embeds itself in the belly of a pregnant woman and emerges holding a puppetised foetus. This, then, is a spatiality of integration, which might involve unidirectional or bidirectional processes of merging puppet and human qualities (intercorporation).

The progression from attitude/affinity on the left to body on the right can thus be seen as a strengthening or magnifying of human–puppet convergence and more substantial human-to-puppet transformation, as a positive attitude towards puppets might be in evidence without any further transformation from humans to puppets. However, this progression is not inevitable, as bodily merging might feature without any indication of a positive attitude towards or personal affinity with puppets on the part of the human being transformed (e.g., if they are possessed against their will). Similarly, the approaches within each category should not be seen as mutually exclusive, as in the case of the interaction between the familial and functional themes, but – consistent with the previous section – should be considered as complementary approaches. Figure 7.1, then, can be considered to provide more detail on the role of constructive complementarities in relation to the form being transformed in addition to those related to character and plot that were identified earlier. Equally, these approaches should not be assumed to run in a direct causal chain from left to right above and below the arrow, as – for example – a positive attitude towards puppets might (or might not) lead to behaviour *through* puppets as much as to behaviour *as* a puppet. Consequently, rather than being a formal process diagram, Figure 7.1 is best considered a patchwork of approaches to the construction of becoming puppet in the works analysed, the varied employment of which leads to unique composite human–puppet spatialities for each tale, which are potentially complex.

## Trans-corporeal borderscapings

The two analyses presented so far in this chapter indicate the need to consider approaches to human-to-puppet transformations beyond physical bodily borders, as facets of attitude and affinity, identity, psychology and affect have come to the fore. Within the broad and diverse approaches explored thus far, we can identify ten specific devices that are employed across these works by which humans become puppets. In brief, these are:

1.  Internalising, for example, of a human spirit in a puppet body or a formerly puppetised spirit in a human body.
2.  Funtionalising, for example, incorporating a puppet mechanism into a human bodily form.
3.  Familiarising, for example, establishing familial relations with or through puppets.
4.  Subjectivising, for example, taking on the identity of a puppet character.
5.  Contextualising, for example, taking on or replicating the qualities or events of a known puppet play.
6.  Vocalising, for example, through the puppet controlling or replacing the voice or speech of the human.
7.  Intercorporeality, for example, through bodily merging or integration.
8.  Behaviourising, for example, enacting behaviours either as or through a puppet.
9.  Accessorising, for example, attaching to a puppet as an accessory or outer layer.
10. Unionising, for example, identifying and empathising with puppets.

This list might be surprising, given the emphasis throughout this volume on intercorporeality and the crossing of human/puppet *bodily* borders, as intercorporeality is now only one of ten devices or tactics used in constructing human-to-puppet transformation. However, as many of these devices still relate to the body, this list is more about adding nuance to the ideas of intercorporeality and borderscaping practices across the bodily border and providing more-than-bodily context for those practices than it is about unsettling the foregoing analyses. It sets the notion of intercorporeality or intercorporation across the human/puppet bodily border established earlier within a broader context, bringing into consideration issues of psychology and identity markers, and the contextual significance of puppet plays and specific puppet characters. One thing remaining to be considered, however, is how we might conceptualise the relations between these varied devices.

As established earlier in this volume, intercorporeality is understood as the merging of bodily forms and materialities across the human/puppet bodily border, whereby that bodily border is blurred, enabling the creation of peculiar – unique and strange – beings and spaces, often but not necessarily resulting in a body grotesque. In addition to this bodily focused form of borderscaping, we can now also discern non-bodily but individualised mergings of puppet and human entities, such as identities, voices, spirits or cognitive capabilities, which fit very well with earlier ideas of the character contorted. Although they are principally grounded in body-based borderscapings (e.g., possession), this is not inevitable and such mergings can be strongly psycho-social in nature (e.g., involving a battle between identities). While these cross the human/puppet border, they are primarily human rather than puppet oriented, for example in the case of human cognitive decline due to dementia or depression. As an additional factor in the construction of puppets beyond the bodily form, these might be conceptualised as epicorporeal, in supplementing the intercorporeal mergings identified in previous chapters, yet despite their supplementary nature they might or might not inform those corporeal borderscaping practices (intercorporation).

Simultaneously, we have also seen how these bodily and non-bodily border crossings can be situated within a broader narrative, performative or historical context, which sets parameters for what forms of bodily and non-bodily borderscaping are appropriate and how they might be expressed. While these also cross the human/puppet border, they are primarily puppet-oriented rather than human-oriented, as in the case of the original Punch and Judy puppet play setting the stage for human protagonists in a new production becoming Punch (or Judy). These might be conceptualised as pericorporeal in acknowledgement of their role in aligning the specifics of the human-to-puppet transformation with the narrative, performative or historical setting selected by the work, and while they might link to bodily (intercorporeal) or psychological (epicorporeal) border-crossings, they are distinct from them as they might involve setting the context for action rather than entailing bodily or behavioural change.

As we have seen, these intercorporeal, epicorporeal and pericorporeal devices can be used individually or collectively and can take diverse forms. Consequently, and as was the case with Pinocchio in Chapter 6, these sit within an even broader ontological space that incorporates each and every possible device. Thus, these constructions of puppets involve multiple corpo-realities and more-than-corporeal elements, which are graduated in the degree to which they relate to either human (or animal) or puppet bodily forms, materials and capacities. For example, they might prioritise the more-than-bodily over the bodily, or they might use a puppet play context to deter-mine the balance between bodily and non-bodily aspects of the process of becoming puppet in either a contradictory or complementary manner. These multiple corporealities, then, all fall within a broad pancorporeal ontological space, which is both corporeal and more-than-corporeal and which incorpo-rates everything from the specifics of the bodies involved to the historical and narrative context of established puppet plays.

This conceptualisation can be presented diagrammatically as in Figure 7.2. Intercorporeality is found to sit alongside epicorporeal devices, framed as non-bodily but individual and human-oriented coming-togethers of human and puppet, for example, identity. Intercorporeal and epicorporeal devices sit within a broader pericorporeal setting, framed as puppet-oriented relational or contextual means of establishing puppet–human coming-togethers, for example, within the context of the Punch and Judy show as a core narrative device. These diverse yet related devices are found to be employed variously in isolation or in combination, thereby delimiting the borderscape for any given work. More broadly, however, we can conceive of a pancorporeal onto-logical space of human becoming puppet, and – by extension – puppet becoming human, constituted as all possible border-crossings.

Many of the ten devices listed above map neatly onto these corporealities, for example familiarising and unionising fall under the heading of epicorpo-reality, accessorising and functionalising are examples of intercorporeality and contextualising is pericorporeality. However, others are more ambiguous as they could be either intercorporeal or epicorporeal or indeed both (e.g., internalising), while specific devices might also combine approaches, for example, using possession to control voice, and any device could be informed by a pericorporeal context, but is not necessarily so informed. We can, then, think in either broad categorical terms about the approaches taken to puppet

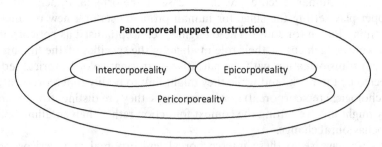

*Figure 7.2* The multicorporeality of puppet-becomings.
Source: Author.

construction as a means of characterising the borderscape that underpins it, or in more detail about the specific devices employed that define and unsettle the borders at its heart, and while there are relations between these two, those relations are neither definitive nor static, thereby both enabling and accommodating vast diversity in the construction of puppets.

Extending this conceptual work further, we can think about these devices and corporealities as a series of ecologies and technologies, with ecologies encapsulating those that are more human-oriented and technologies those that are more puppet-oriented. This is, however, something of a crude distinction. The ecologies of becoming-puppet, for example, encompass the relational aspects, including familiarity and sociality and their associated devices (e.g., subjective, familial), but the pericorporeal space also functions ecologically in setting the habitable context for that specific work in accordance with a pre-existing puppet character or plot, yet this epicorporeal space was associated with puppets, which should make it a technology rather than an ecology. Moreover, the pancorporeal space can also be considered an ecology as it constitutes every possible means by which this puppet might be constructed, as can the borderscape as it sets the parameters for the contradictory and complementary relations that contribute to the entity that becomes. Similarly, the technologies of becoming-puppet encompass the material aspects and associated devices of puppet construction (e.g., functionalist, intercorporeal), but the devices themselves – whether ecological or technological – can also be considered technologies as they are a means by which construction is achieved. Moreover, both the pancorporeal space and the borderscape can be considered technologies as they are similarly a means by which construction is achieved. Consequently, while we can categorise different devices and perceive a certain hierarchical relationship between the multicorporeal spaces, thinking about these diverse features of becoming-puppet in terms of interwoven ecologies and technologies emphasises the ambiguity central to the puppet as neither ecology nor technology is entirely discrete. Conceptualising the construction of puppets in this way thereby embodies the ambiguity of the puppet's physical form and ontological status in its conceptualisation and is both reminiscent and illustrative of the reversal in understandings (see Chapter 1) of the grotesque from a concept without a form (Harpham 2006) to a form that escapes conceptualisation.

Thinking in this way also enables us to forge links between the different conceptualisations developed through this volume, such as the multiple corporealities above, the character contorted and the borderscape. There is, however, no direct correspondence between ecology/technology and either inter-/epi-corporeality or the body grotesque/character contorted, but a host of potential intersections between them, as these different corporealities and grotesqueries can be constructed through ecological or technological (more human or more puppet oriented) devices. While there is a degree of correspondence between intercorporeality and the body grotesque and between epicorporeality and the character contorted, they are neither mutually exclusive nor bound together, so ecology/technology and inter-/epi-corporeality are best considered intersecting axes of puppet construction, which might or might not be set within a pericorporeal space or context in constituting the borderscape, and which give rise to different configurations of body grotesque and character contorted. Similarly, while both the pancorporeal space and the borderscape are simultaneously ecology and technology, they are not the same thing, as the pancorporeal space consists of every possibility for the construction of any particular puppet but the borderscape potentialises certain of these possibilities for enactment, with those potentialities being enacted (or not) through the specific border-crossing devices selected. Through this conceptual discussion, then, we reach a point at which we can generate a definition of borderscaping beyond that outlined in Chapter 3 and a diagram (see Figure 7.3) of the borderscape in cultural constructions of puppets.

Consistent with more conventional state-based articulations of a borderscape (see, for example, dell'agnese and Amilhat Szary 2015; van Houtum and Eker 2015; Robson 2019), this borderscape is an area around a border but the border is that between human and puppet rather than between states, and the area is less a spatial extent in a measurable sense than an ontological field of potentiality. Different elements of this field of potentiality are brought together in either contradictory or complementary fashion, through a series of ecological or technological devices (e.g., familiarising and functionalising) that are oriented around intercorporeal (bodily) and/or epicorporeal (more-than-bodily) border-crossings. We can consider the different approaches to the border identified in Chapter 5 – theatrical/radical dissolution or strengthening of the border – as borderscaping strategies, which accentuate or mask the human/puppet border, and this combinatorial work – through these varied devices and strategies – is what constitutes borderscaping. It is through this borderscaping that both puppets and their peculiar worlds are constructed, which might or might not be informed by a

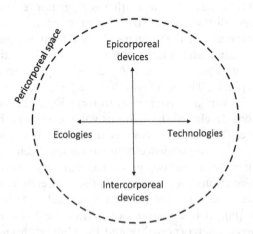

*Figure 7.3* The borderscape.

Source: Author.

pre-existing pericorporeal space, and which might or might not create a body grotesque and/or a character contorted. Ultimately, then, despite the expansive array of different conceptualisations unearthed through the varied analyses in these chapters, they fit together to provide a substantial and organised account of the construction of puppets in popular culture.

Given this conceptualised intersplicing of human and puppet attributes and capacities, and to draw this chapter to a close, it is worth considering how we can make disciplinary use of the puppet's and the grotesque's generativity in interrogating what it is to be human. The grotesque has been characterised as providing a critical route into the human condition at both the level of the body (grotesque) and the level of society (the political potential of the carnival) (Bakhtin 1984; Danow 1995; Li 2009; Connelly 2012; Pilný 2016). Puppets have similarly been attributed with the ability to draw attention to humanness and human qualities as humans recognise themselves both more truly and as more strange in relation to the puppet (Gross 2011; Francis 2012). Lacking an interior world or life, the puppet undermines attachment to subjectivity and traditional roles of subject and object are reversed as it is the object that now entices interrogation of humanity (Zamir 2010; Francis 2012). Indeed, in twentieth-century theatre practice, the puppet was claimed as a theatrical subject (Francis 2012), swerving away from any essentialised notion of the subject in a manner entirely consistent with contemporary disciplinary concerns with how we can conceptualise human subjectivity within more-than-human sensitivities for distributed and performative subjectivities enrolling both human and nonhuman elements (Thrift et al. 2010). While intercorporeality is a firm feature of the current volume given its centrality to the constructions and uses of puppets, this chapter has provided specificity within the general idea of corporeality as including both percepts and affects (Wylie 2002) by distinguishing between intercorporeality as bodily and epicorporeality as psychological (both cognitive and pre-reflective) but also recognising that the psychological and the physiological are intricately interwoven. Setting these two approaches within the context of a pericorporeal grounding (through play or character or historical context) helps to counterbalance our interest in the individual, the body or the event with sustained attention on the situatedness, contingency and constrained agency that characterise them, thereby providing a potential means to calibrate the unique, the distributed and the co-constituted aspects of subjectivity beyond puppet constructions by considering puppets – in this instance – as a proxy for (more-than-)humans. An important distinction needs to be drawn here, however, between the use of puppets as proxies in a performative sense, which in Chapter 3 was considered to hinder the puppet's performativity by precluding interrogation of puppetness, and similar use of puppets as proxies in a theoretical sense, which is here considered helpful in pluralising, distributing and dehumanising the subject.

Significantly, calls have been issued for the establishment of a concept that can replace the subject with a collective subjectivity (Thrift et al. 2010), and in this context, the puppet perhaps comes to the fore as a different kind of nonhuman vocabulary for what might otherwise be called subjectivity, which is founded upon the bridging of diverse topological gulfs between conceptual and ontological categories. In being simultaneously both subject and object, singular and multiple, human and nonhuman, agentive and not, vital and inert, perhaps rather than subjectivity, there is potential in the notion of 'puppetivity' in accommodating the blended, parallax and ambiguous nature and status of the simultaneously more-than/less-than-human, by focusing first and foremost not on the individualistic idea of the subject nor on the non-specific generalised more-than-human, but on the inherent ambiguity upon which the puppet rests. In this theoretical context, the borderscape is less an ontological sphere of intentional border-crossings in the construction of specific puppets and more an ontological sphere of topologically unified contradictions within, from and through which incidental border-crossings give rise to contingent provisional puppetivities that are concurrently human and nonhuman, subject and object, singular and multiple. On this reading, borderscapings are performative eventualities within a delimited and situated potentiality (the borderscape) that provides the conditions of possibility for the becoming of unique puppetivities that are always-already both more-than and less-than human. Of relevance to substantial sub-disciplinary terrain from more-than-human geographies to psycho-geographies, this multicorporeality brings new perspectives and opportunities for the exploration and extension of disciplinary engagements with both (inter-)corporeality and (inter-)subjectivity and positions the puppet as a valuable member of the geographical community.

## Conclusion

This chapter complements Chapter 6 by delivering an analysis of the transformation from humans to puppets in varied cultural works, including some not previously analysed. The chapter began with a comparative analysis of these works in relation to the understandings of puppets that had emerged through the preceding chapters, which identified different perspectives on aspects of the dimensional framework, constructive contradictions and ideas of borderscaping. While the dimensional framework developed in Part 2 was found to be largely applicable to the works analysed here, additional features of puppet construction and human-to-puppet transformation were identified including the morphing of bodily capacities, the significance of pre-existing puppet plots and self-actualisation, providing new perspectives on the contradictory construction of puppets by introducing a suite of constructive complementarities that work alongside constructive contradictions in the construction of puppets.

Subsequently, and extending the previous discussion, the implications of this analysis were employed to develop a diagram of different approaches to the construction of the human-to-puppet transformation, which fell into five categories – perspective, behaviour, focus, identity and body – and were framed as paired complementarities. The progression from left to right (see

Figure 7.1) was presented as a strengthening or magnifying of human–puppet convergence and more substantial human-to-puppet transformation, but this progression is not inevitable, as categories might be used in isolation, giving rise to the description of this diagram as a 'patchwork' of approaches to becoming-puppet, each of which has its own spatiality and through their varied combinations generate unique and potentially complex spatialities within each work.

Thirdly, I interrogated these approaches further and unearthed ten specific devices that are employed to transform humans into puppets, which provided both more detail and a broader context for the foregoing emphasis on intercorporeality. The differences and relations between these devices were used to conceptualise a suite of corporealities that work either in isolation or in combination to configure these diverse human-to-puppet transformations within an overarching pancorporeal ontological space. Subsequently, these were re-articulated as a series of ecologies and technologies, which were used to integrate the different conceptualisations established through the preceding chapters and to formulate both verbal and diagrammatic descriptions of the borderscape and borderscaping. Finally, the potential implications of this interrogation of puppetness for disciplinary concerns about the status of the subject were explored, with the proposed use of puppets as theoretical (not performative) proxies for humans to enable the conceptualisation of a simultaneously more-than/less-than single-multiple subject–object entity and the possible utility of puppetivity as a substitute term for both singular and collective subjectivities.

This chapter, then, both affirms and refines the understandings of puppet uses and spatialities that emerged through the previous chapters. It generates a specific and detailed account of how human-to-puppet transformations are constructed and the spatialities that they generate, proposing a set of constructive complementarities with respect to character, plot and form, a nested suite of corporealities and a fuller and more formal articulation of the borderscape as substantial conceptual contributions. It also brings to an integrative conclusion the varied analyses on which these findings and contributions are based and offers a possible way forward with respect to critical disciplinary engagements with subjectivity, so all that remains to be done is to pull together and flesh out the breadth and depth of contributions across this volume in the context of the construction of puppets, the grotesque and the discipline of geography, which is the task of the conclusion.

## References

Bakhtin, M. (1984) *Rabelais and his world*. Indiana University Press: Bloomington, IN.

Connelly, F.S. (2012) *The grotesque in Western art and culture: The image at play*. Cambridge University Press: New York.

Danow, D. (1995) *The spirit of carnival: Magical realism and the grotesque*. The University Press of Kentucky: Lexington, KY.

dell'agnese, E. & Amilhat Szary, A.L. (2015) Borderscapes: From border landscapes to border aesthetics. *Geopolitics*, 20: 4–13.

Francis, P. (2012) *Puppetry: A reader in theatre practice*. Palgrave Macmillan: Basingstoke.

Gross, K. (2011) *Puppet: An essay on uncanny life*. University of Chicago Press: Chicago, IL.

Harpham, G. (2006) *On the grotesque: Strategies of contradiction in art and literature*. The Davies Group Publishers: Aurora, CO.

van Houtum, H. & Eker, M. (2015) Borderscapes: Redesigning the borderline. *Territorio*, 72: 101–108.

Li, M.O. (2009) *Ambiguous bodies: Reading the grotesque in Japanese Setsuwa Tales*. Stanford University Press: Stanford, CA.

Pilný, O. (2016) *The grotesque in contemporary Anglophone drama*. Palgrave Macmillan: London.

Robson, M. (2019) Metaphor and irony in the constitution of the UK borders: An assessment of the 'Mac' cartoons in the Daily Mail newspaper. *Political Geography*, 71: 115–125.

Thrift, N., Harrison, P. & Anderson, B. (2010) "The 27th Letter": An interview with Nigel Thrift. In: B. Anderson & P. Harrison (eds) *Taking place: Non-representational theories and geography*. Ashgate Publishing Ltd: Farnham and Burlington, 183–198.

Wylie, J. (2002) Becoming-icy: Scott and Amundsen's South Polar voyages, 1910–1913. *Cultural Geographies*, 9: 249–265.

Zamir, T. (2010) Puppets. *Critical Inquiry*, 36(3): 386–409.

# Conclusion to Part 3

This third and final part has attended to the becomings – transformations – from puppets to humans and from humans to puppets, as constructed in various cultural works. Chapter 6 focused on Pinocchio and conducted a mainly visual analysis of four illustrated tellings of the tale to unpack the becoming-human of the titular puppet, while Chapter 7 examined the becoming-puppet of various human characters in a set of diverse works. In this conclusion, I briefly recap on the findings of each chapter through a comparative discussion, before bringing the outcomes of these chapters together as a step towards formulating a composite diagram of ecologies and technologies of puppet-becomings to provide an answer to the question of how puppets (and their varied becomings) are constructed. Subsequently, I flesh out in more detail the disciplinary implications of this part in relation to geocriticism and posthumanism, through which discussion I reaffirm the potential of using puppets as proxies for humans in a theoretical context as a means of furthering posthumanist debates concerning the human subject.

The analysis in Chapter 6 identified a series of contradictions by which Pinocchio is variously constructed, which were used to elaborate on Carlo Collodi's own appreciation of the 'but' of the puppet to speak to their inherent contrariness. Materially, these contradictions revolved around flesh/wood, form/expression, body/head, body/mind, and – while consistent with the findings in earlier chapters – supplemented these previous findings. Exploring Pinocchio's sociality unearthed both a social dimension on which Pinocchio is variously and unstably constructed in relation to humans, animals and objects, and an affective dimension focused on threat, on which the source and target of the threat might or might not be identifiable. While Pinocchio's metamorphoses were surprisingly under-represented in the imagery in the works examined, their analysis identified two further sets of constructive contradictions, one relating to the treatment of the ambiguity in Pinocchio's ontological status at the end of the story, and the other relating to in/consistency between the narrative and illustrative versions of the tale in a single work. Both were considered to have implications for the integrity of the work and for the continuing becoming of Pinocchio as an intertextual, intersemiotic mythological figure. A distinction was drawn between the multiple borderscapes at play in each version of the tale and the micro-spatialities

DOI: 10.4324/9781003214861-17

that emerge from those borderscapes, which vary as Pinocchio's materiality and sociality shift, and the significance to the integrity of the tale of consistency between these spatialities was highlighted. Ultimately, Pinocchio was confirmed as going through multiple transformations through these shifting borderscapes and micro-spatialities even in versions that do not result in him becoming a real boy.

In Chapter 7, I presented an analysis of human-to-puppet becomings in various works, which established both the applicability of previously identified constructive contradictions and certain refinements to those understandings to accommodate the additional mechanisms by which these transformations were constructed, over and above how puppets are constructed as pre-existing entities. Moreover, a supplementary set of constructive complementarities was unearthed, which operate in relation to character, plot and form, establishing a somewhat different approach to the construction of puppet becomings compared to the construction of puppets as contradictory entities. A schematic of such approaches adopted towards the transformation of humans into puppets was presented and described as a 'patchwork' continuum from humans to puppets. Through this analysis, ten specific devices were identified that are employed to transform humans into puppets, which were collectively conceptualised as a suite of corporealities that work either in isolation or in combination within a pancorporeal ontological space of becoming-puppet. The chapter highlighted that – just as we can distinguish between the border and the borderscape – so we can distinguish between the broader categorical approach to constructing it and the specific border-crossing devices that define and unsettle the borders at its heart. The chapter concluded by integrating ideas of the borderscape with the various corporeal spaces, the ecologies and technologies, and the body grotesque and character contorted and by presenting a fuller elaboration of both the borderscape and borderscaping.

Each chapter, then, affirms the utility of the understandings of puppets that emerged from the preceding chapters and is informative in its own way in contributing to the development and refinement of these understandings. Addressing a specific direction of metamorphosis in each chapter, though, risks generating contradictory, conflated or confused accounts of puppet-becomings when considered together, so the next section examines the relevance of the analyses in Chapters 6 and 7 to the constructions and metamorphoses explored in the other chapter. Through this integrative discussion, a further composite framework presenting the varied ecologies and technologies through which puppets and puppet-becomings are constructed is proposed. Finally, the significance of this work for geography – both cultural and otherwise – is explicated.

## Ecologies and technologies of puppet-becomings

The starting point for this discussion is consideration of the applicability of Figure 6.1 beyond the confines of the visual analysis through which it was

generated. With respect to material contradictions, the various border crossings identified in relation to Pinocchio fit well with those unearthed in
Chapter 7, albeit that the latter were more varied, and the continuum from
individual to social puppet relations was also consistent between the chapters. Animal relationality is less evident in Chapter 7, although this is to be
expected given the specific cultural context of Pinocchio, suggesting that no
change is required to Figure 6.1, merely acknowledgement that different
aspects of it will feature to different degrees in different contexts.

Similarly, the affective dimension from Chapter 6 does not feature explicitly in the table of dimensions in Chapter 7, although diverse affective tenors
were evident, including fear (*Dead Silence*), anxiety (*Robert*), terror (*Dolls*),
but also excitement/exhilaration (*Being John Malkovich*), melancholy (*The
Beaver*) and love (*The Puppet Maker's Bones*). Consistent with Figure 6.1,
where there is a direct threat identifiable in the works, that threat does generally originate from an individual – whether human, puppet or spirit (*Dead
Silence, Robert, Magicians of Caprona*) – and the target of that threat does
tend to be more collective or multiple in nature, extending to family, friends
and members of the public (*Dolls, Robert, Rivers of London*), although there
are also examples that lack specific threat (*The Beaver; Being John Malkovich*).
Again, such variation is not so much problematic as an indication that affect
can be – and is – employed in different ways across the diagram, rather than
constituting a separate approach, whereas the narrow affective focus in
Chapter 6 arose from the examination of specific images depicting individual
episodes in Pinocchio's adventures.

The consistency between illustrative and narrative versions of Pinocchio
finds alternative expression in Chapter 7 in the context of the parameters set
for the human-to-puppet transformation by the play context of the Punch
and Judy story, bringing us back to questions as to the role that the original
story should play in the borderscape for its subsequent reformulation.
Ambiguity is perhaps more challenging to accommodate in relation to
Chapter 7, as this issue acquires particular significance in the context of
Pinocchio given that the ambiguity of his ontological status is fundamental
to the tale, whereas the specific purpose of the transformations in Chapter 7
is to establish a clear transition from one status to another. However, ambiguity is a core feature of many of these latter transformational tales, for example
in building a sense of anticipation in advance of the 'big reveal' that the
human is in fact a puppet. Similarly, while not ubiquitous in the human-to-
puppet transformations, it is also not uncommon for uncertainty to persist
beyond the end of the story or film as to whether potential remains for further transformations to occur (*Rivers of London, Dead Silence*) or for the way
to be paved for a recurrence of the events that have unfolded (*Robert, Dolls*),
so again, the findings from Chapter 6 fit well with those of Chapter 7, requiring only mindfulness as to the greater specificity in the former and greater
variety in the latter.

Exploring now the applicability of the findings of Chapter 7 to those of
Chapter 6, similar translatability is evident. Considering the 'patchwork'

continuum first (Figure 7.1), in terms of perspective, Pinocchio exhibits both attitude and affinity, although – as would be expected given his aspirations to become human – this attitude and affinity is as much towards humans as towards puppets. It is in relation to identity and body that the findings of Chapter 7 become less applicable, although this is due to distinct differences between Pinocchio and the other works examined. In Chapter 7, puppet and human typically start out as separate entities but become one while often containing two identities, whereas Pinocchio is both human (e.g., emotional) and puppet (e.g., wooden) to start with, so rather than a merging of bodily forms and identities he leaves behind his puppet form when he takes on his full humanity, but he has the same identity despite his bodily changes. However, the label 'morphing' already encapsulates diverse approaches to the modification of the human–puppet 'body', so while Pinocchio's transformations are very different to those in Chapter 7, they can still be accommodated by the flexibility in the 'patchwork' diagram.

Finally, in relation to the ten devices identified in Chapter 7, these also apply to Pinocchio but in specific ways. Internalising, familiarising and behaviourising bear resemblance to the transformation of Pinocchio as he already has aspects of the human within himself (although – as above – this would be better termed externalising in that he extracts his humanity from his puppet form), he has clear human familial relations, and his behaviours are already identifiable as human. Moreover, the subjectivising and contextualising devices are highly applicable because Pinocchio sets the context for his own retelling. While it is perhaps unlikely that any single puppet construction will entail each of these ten devices, rendering it unproblematic that Pinocchio does not exhibit them all, the devices can be supplemented and/or rephrased as follows to accommodate these differences:

1. **Internalist**, for example, the end state of the transformation is internal to the transforming entity, whether through incorporation or extrication.
2. **Funtionalist**, for example, incorporating human/puppet mechanisms and uses into the form of the other.
3. **Familiar**, for example, establishing familial relations with or through the entity/ies into which a character is transforming.
4. **Subjective**, for example, taking on the identity of a specific puppet/ human character.
5. **Contextual**, for example, taking on or replicating the qualities or events of pre-existing narrative frameworks (such as a puppet play or human historical event).
6. **Vocal**, for example, the puppet controlling or replacing the voice/speech of the human.
7. **Intercorporeal**, for example, through bodily merging or switching.
8. **Behavioural**, for example, enacting behaviours either as or through that into which a character is transforming.
9. **Accessorising**, for example, attaching to a puppet/human as an accessory, clothing or outer layer.

10. **Unionising**, for example, identifying and empathising with that into which a character is transforming.

The final diagram in Chapter 7 – of the pancorporeal space – remains as it is, as there are no substantive changes to the devices or their inter-relations. However, the task still remains to integrate the varied findings of Chapters 6 and 7, which is presented diagrammatically in Figure C3.1.

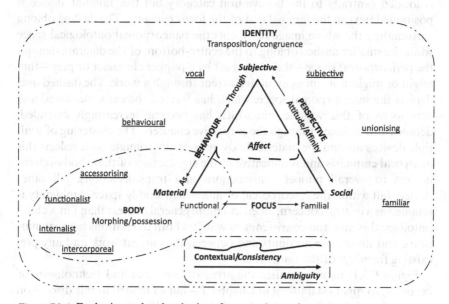

*Figure C3.1* Ecologies and technologies of puppet-becomings.
Source: Author.

**Key:**

| | |
|---|---|
| *Bold italics*: | dimensions/continua from Chapter 6 (becoming-human) |
| **BOLD CAPITALS**: | perspectives from Chapter 7 (becoming-puppet) |
| Underlined: | devices from Chapter 7 (becoming-puppet) |
| ------- : | groups of devices/dimensions |

In this diagram, the triangle of contradictory continua (from Chapter 6) is evident and affect remains distributed across this element, although without the specifics of the Pinocchio analysis for the sake of presentational parsi-mony. The five categories of the 'patchwork' diagram (from Chapter 7) are incorporated (indicated in bold capitals), with three of them mapping neatly onto the edges of the triangle, for example, the behavioural category spans from material, denoting a human character behaving like a puppet, to the subjective, denoting a human character behaving through puppets. The other two 'patchwork' categories are more tightly associated with apices of the triangle, for example, identity is anchored on 'subjective'. The ten devices are represented by the underlined terms, and these are located in relation to different aspects of the central structure. For example, the 'behavioural' device is located centrally to the behavioural category but the 'familial' device is positioned next to the familial end of the focus category. The dashed oblong surrounding the whole image represents the pancorporeal ontological space while the smaller dashed oblong to the centre-bottom of the diagram denotes the pericorporeal space – the context set by a puppet character or play – that might or might not run as an undercurrent through a work. The dashed oval depicts the intercorporeal space, which has formed the early, sustained and core focus of this volume, but which has become increasingly extended, detailed and nuanced through the successive chapters. The clustering of multiple devices around the material/body apex of the triangle both reflects this analytical emphasis and is indicative of the significance of these body-related devices to overall puppet constructions and transformations. All other devices fall within the epicorporeal or more-than-bodily space. Ambiguity is retained as a central concern, albeit in a more general context than Pinocchio's ontological status, and consistency now refers both to relations between narrative and illustrative accounts and between the current work and any pre-existing framings of the same story.

Figure C3.1, then, organises the array of ecologies and technologies of becoming-puppet established in Chapter 7 – but as refined in this discussion – around a central structure that pulls together the range of contradictory continua and diverse approaches to the construction of Pinocchio's becoming-human identified in Chapter 6. In doing so, it specifies how these collective puppet-becomings have been constructed and – in their consistency with previous findings – also incorporate those earlier insights, providing a schematic answer to the guiding question of how puppets are constructed. It both brings together and distinguishes between the intercorporeal devices that are most strongly associated with the creation of the body grotesque and the epicorporeal devices that bring their own opportunities for the creation of the character contorted. It also accommodates both the contradictory and complementary approaches to borderscaping without specifying either one and provides for the possible but not essential intertextual influence of prior puppet characters and stories. Figure C3.1 can thus be considered a constructional framework for the borderscaping of puppets.

## Puppet geographies

The analyses presented in Part 3 hold significance for geography and beyond in two primary ways, relating to geocriticism and subjectivity, with the latter here being considered in the context of posthumanism. As a starting point, they highlight the significant spatial character of the different approaches adopted to constructing puppets and their becomings, which is reminiscent of the attitudes and orientations of the puppet masters to their victims identified in Chapter 2. Moreover, the analysis of Pinocchio signposted the implication of the spatialising aspects of the tales for the integrity of the tale being told and the character being constructed as a micro-spatiality of the social might be inconsistent with a corresponding micro-spatiality of the material. While these spatialities remain implicit in the works analysed, acknowledging both their presence and their potential impact on the finished work draws attention to their availability for both more targeted literary analysis and critique, and for more strategically spatialised approaches to writing, creative and otherwise. In terms of literary analysis, charting these spatialities provides an alternative lens through which to engage with a work that takes us beyond the explicit content to consider the spaces generated by it, irrespective of any real place/s to which it might relate and any fictional or conceptual spaces that it might narrate.

Significantly, this distinction constitutes a redefinition – or an extended definition – of geocriticism as rather than seeking a critical understanding of how a singular and identifiable 'real' place is constructed through the analysis of the explicit content of one or more literary work (Westphal 2011), this form of geocriticism would use the analysis of the constitutive spatialities within a literary work to interrogate and critique the work. Here, the text is not understood as standing for 'the' place nor as emplacing us within it (Moslund 2011), although it might do these things, but space is rather understood in terms more akin to Tally's (2011) view of geocriticism as attending to the spatial practices involved in literature. However, the emphasis here would be less on the spaces that are explicitly written about (e.g., a setting) and more on the spaces that are implicitly written into being while *not being* written about. Unlike established ideas of geocriticism, the emphasis is on neither the representational content nor the referential force of a literary work (Prieto 2011) but on the non-represented spatialities constituted by the representational content that have constitutive force with respect to the integrity of that content but that are commonly overlooked. Both the 'geo-' and the 'critical' aspect of geocriticism is thus redirected: The 'geo-' refers to the spatialities that are implicit within the work rather than being explicit either within or outside the work, and the critique is directed towards the work itself rather than any place that it might happen to be about.

Chapter 7 related the interrogation of puppetness in the construction of human-to-puppet transformations to critical engagement with the human condition, consistent with the notion of both puppets and the grotesque as questioning this condition, suggesting that puppetivity might provide an alternative to subjectivity in encapsulating an always-already more-than/less-than-human entity. This discussion is extended here in the context of disciplinary debates about the need for or perils associated with a posthuman resurgence of humanism. As interest in the geographies of nonhumans has blossomed over the past few decades (e.g., with the growth of animal geographies), so too have efforts to decentralise the human from geographical enquiry and practice (Braun 2004). For example, multinaturalism as a mode of environmental governance holds all species and their needs as equal in their role within the determination of the environmental future and seeks inter-species conviviality rather than prioritising the political status of the human (Lorimer 2012). However, such disciplinary developments have prompted resultant concerns about the need to maintain minimal humanism (Braun 2004; Simpson 2017) in recognition of human distinctiveness. For example, recent work has proposed a critical phenomenology and a new humanism that would allow for retaining consideration of human uniqueness without putting the human back on its pedestal (Simonsen 2013; Ferretti 2020), in efforts to separate human distinctiveness from human exceptionalism. Similarly, other voices caution that such a move would be at odds with the original efforts to decentre the human (Williams et al. 2019), leaving the matter far from settled. Indeed, questions have been posed as to what is left of the human and what comes after the human if we wander too far down the posthumanist path towards subjectivity being constituted in asocial practices and distributed among objects and technologies (Simpson 2017).

Of interest here is the similarity between these posthumanist debates and the construction of puppets through the interrogation of puppetness, as they both render different conceptual categories equivalent to varying degrees. In each case, the question is essentially the same: How different is the human to the comparator, whether that comparator be a puppet or an animal? Puppets, then, are a posthuman phenomenon and – in both their variety and their grotesquery – they provide an ontological category, material form and embodied practice through which to pose critical and philosophical questions as to what it is to be human (or nonhuman). The puppet provides a more-than/less-than-human figure through which we can rethink some of these issues, as a conjoined entity in which both parts are equal and essential yet distinct and unique. Importantly, neither part holds meaning in isolation as puppetness – their conjunction – is determined by use, not latency (Francis 2012), so it is that very conjunction that comes to the fore in considering subjectivity (or puppetivity) in a manner entirely consistent with contemporary emphasis on the performative nature of subjectivity. The human, while part of that conjunction, is only ever part of it, and in isolation becomes inconsequential. Potentially, then, puppets enable us to avoid the risk of falling back on assumptions of human exceptionalism by sustaining our focus on the human's multi-corporeal and more-than-corporeal interconnectedness and simultaneously attending to the human's base materiality as much as its particular capacities. Playing at the limits of the co-ordinates of human experience and capacity, the puppet opens the door to alternative ontological universes and processes of becoming – as has been suggested in the context of Guattarian philosophy and science fiction (Hynes and Sharpe 2021) – but without the need for either science fiction or straying beyond the human/ other framing of subjectivity.

Thus, if we consider there to be productive potential in seeing puppets as theoretical (not performative) proxies for humans in the sense that puppetivity might be a candidate replacement for subjectivity (see Chapter 7), then the answer to the two questions posed above – what is left of the human and what comes after the human if we go too far down the posthumanist path? (Simpson 2017) – is the same: The puppet. Paradoxically, though, this would not leave the human out of the picture because the human is part-and-parcel of the puppet: The figure of the puppet is incomplete without its human part but in the context of the puppet, the human as a discrete (let alone exceptional) entity makes no sense. Rather than thinking about the human as constituted through its relations with other things, here we are thinking about how the human contributes to the constitution of other things, not as a creator but as a component, so the human is important and distinct but not dominant and certainly not exceptional. The puppet, therefore, holds potential with respect to the establishment of a posthuman humanism, preventing us from going *too far* down the posthumanist path by enabling us to decentre but not jettison the human, simultaneously keeping the distinctiveness of the human in focus and out of sight. In essence, then, and ultimately, we now have a (theoretical) theatrical dissolution of the human/puppet border insofar as we dissolve the distinction between human and puppet even while accentuating our focus on human/puppet distinctiveness, potentially accommodating a minimal humanism within a posthumanist geography through the proxy of the puppet and the replacement of subjectivity with puppetivity.

## Conclusion

Despite delivering two distinct and multifaceted analyses of bi-directional puppet-becomings and despite generating diverse and sometimes seemingly incompatible findings, this conclusion has shown the mutual translatability of these findings and has generated specific contributions to geography and beyond. Moreover, the two sets of findings have been brought together and presented diagrammatically, and have been conceptualised as a suite of interwoven ecologies and technologies of puppet-becomings that themselves reflect the convoluted and contrary nature of the puppet-becomings upon which they have been based.

The visual analysis of Pinocchio in Chapter 6 was developed into an alternative understanding and practice of geocriticism whereby the spatialities analysed are implicit in the work, being written into existence but not being explicitly written about, and the critical attention is directed to the implications of those spatialities for the integrity of the work itself. Intercorporation has been a sustained interest through these final two chapters (and the rest of the volume), although Chapter 7 generated a more comprehensive and detailed appreciation of the varied forms that this can take, being developed into a nested suite of corporealities that take into account both psycho-social modes of puppet-becoming and the significant role often – but not always

– played by pre-existing contextual factors. This was related both to ongoing and growing disciplinary concerns and debates as to the status of the human, as puppets were deemed to provide a material form and embodied practice through which to explore such issues and debates. Finally, this third and final part drew together many of the varied conceptualisations of puppet constructions and becomings from the preceding chapters to present both a fuller and more formalised account of the borderscape and borderscaping and an integrative constructional framework that summarises the varied puppet-becomings examined here.

It is also this integrative analytical and conceptual work that enables us to build on the work of the previous chapters and generate a more detailed response to the question guiding Part 3: How are puppets constructed? The previous chapters have already established the importance of factors such as the interrogation of puppetness and the borderscape but Chapters 6 and 7 extend this understanding in both general and more specific ways. At the more general level, Part 3 pin-points complementarities as much as contradictions in the construction of puppets and – although complementarities did emerge in earlier analyses – they featured more strongly in the construction of puppet-becomings than in the construction of puppets as entities. The significance of implicit spatialities that emerge from puppet narratives rather than being explicitly present in those narratives was highlighted, especially in terms of the potential for discrepancies and inconsistencies to undermine the integrity of both the puppet and the tale. Continuing the interest from Part 2 in the relationship between implicit and explicit registers, this emphasis on implicit rather than explicit spatialities reinforces that puppets are constructed on both registers. More specifically, Part 3 has formulated a series of interwoven technologies and ecologies, which are primarily (although imperfectly) puppet-oriented and human-oriented, respectively, and which revolve around a suite of multicorporealities that incorporate bodily, psychosocial and contextual means of puppet construction. Through these conceptualisations, connections were forged between the multicorporeal technologies and ecologies that emerged in Part 3 and the strategies of borderscaping identified in Part 2, which – together – both further substantiate the puppet as grotesque and constitute specific and diverse puppet bodies grotesque and characters contorted.

Overall, then, puppets are constructed through diverse strategies of borderscaping (dissolving or strengthening the human/puppet border), which variously focus on inter-, epi- and peri-corporeal mergings, which – in turn – include specific ecologies and technologies. They might (or might not) be informed by pre-existing puppet plays and characters and the implicit spatialities that are generated through the construction might be consistent or inconsistent with each other, with the borderscape and between narrative and visual versions of the puppet tale, which can either bolster or undermine the construction. The most peculiar puppets and puppet worlds are created through targeted employment or interrogation of puppetness, and some of the most acute affective impacts are stimulated by stark revelation (radical strengthening) of the human/puppet border as this draws out the interaction and contradiction between implicit and explicit engagements with the puppet, whereby the implicit-explicit duality constitutes yet another topological gulf at the heart of these puppet constructions.

Finally, considering these multicorporealities in the context of disciplinary debates about the future of subjectivity in a posthumanist geography, led to the proposal of an affirmative use of the puppet as a theoretical proxy for the human. Specifically, this took two forms: The potential for 'puppetivity' to function as a replacement for subjectivity and the potential for the puppet to provide an answer to questions as to what replaces and what comes after the human in these posthumanist times. In true grotesque fashion, the puppet and the human constitute each other while being diametrically opposed, such that the replacement of subjectivity by puppetivity need not eliminate either the human or subjectivity from the discipline of geography because the human is part of but not superior to the puppet. In other words, the same 'trick' by which the human/puppet border is theatrically dissolved in the construction of puppets, whereby that border is both denied and highlighted, might be productively utilised in discussions concerning the future status of the human in posthumanism. While the treatment of puppets as proxies for humans was swiftly identified as an impediment to the performative generation of creative puppet spatialities, the same treatment of puppets as proxies for humans in the context of academic debates about subjectivity and posthumanism is reframed as fertile terrain in its capacity to provide a replacement for subjectivity (puppetivity), which – counterintuitively – keeps the human obliquely in focus.

# References

Braun, B. (2004) Querying posthumanisms. *Geoforum*, 35: 269–273.

Ferretti, F. (2020) Traveling in lifeworlds: New perspectives on (post) humanism, situated subjectivities, and agency from a travel diary. *Annals of the American Association of Geographers*, 110(6): 1653–1669.

Francis, P. (2012) *Puppetry: A reader in theatre practice*. Palgrave Macmillan: Basingstoke.

Hynes, M. & Sharpe, S. (2021) Cosmic subjectivity: Guattari and the production of subjective cartographies. *Area*, 53(2): 311–317.

Lorimer, J. (2012) Multinatural geographies for the Anthropocene. *Progress in Human Geography*, 36: 593–612.

Moslund, S.P. (2011) The presencing of place in literature: Toward an embodied topopoetic mode of reading. In: R.T. Tally, Jnr (ed) *Geocritical explorations: Space, place and mapping in literary and cultural studies*. Palgrave Macmillan: New York, 29–43.

Prieto, E. (2011) Geocriticism, geopoetics, geophilosophy and beyond. In: R.T. Tally, Jnr (ed) *Geocritical explorations: Space, place and mapping in literary and cultural studies*. Palgrave Macmillan: New York, 13–27.

Simonsen, K. (2013) In quest of a new humanism: Embodiment, experience and phenomenology as critical geography. *Progress in Human Geography*, 37(1): 10–26.

Simpson, P. (2017) Spacing the subject: Thinking subjectivity after non-representational theory. *Geography Compass*, 11: e12347.

Tally, R.T. Jnr (2011) Introduction: On geocriticism. In: R.T. Tally, Jnr (ed) *Geocritical explorations: Space, place and mapping in literary and cultural studies*. Palgrave Macmillan: New York, 1–9.

Westphal, B. (2011) *Geocriticism: Real and fictional spaces*. Palgrave MacMillan: New York.

Williams, N., Patchett, M., Lapworth, A., Roberts, T. & Keating, T. (2019) Practising posthumanism in geographical research. *Transactions of the Institute of British Geographers*, 44: 637–643.

# Conclusion

Bringing this volume to a close, this conclusion delivers four sets of summative findings. First, I provide a summary of the foregoing discussions, which is organised around the research questions that guided each part and is thus strongly oriented to puppets (what they are, what grants them their performativity and how they are constructed). Next, I progress to thinking about puppets as grotesque to flesh out the conceptual contributions arising from this work for how we think about the grotesque, thereby considering complementarities and rendering congruent as important aspects of puppet grotesquery, distinguishing between the body grotesque and the character contorted, and the spatialisation of the grotesque beyond the grotesque entity itself. Subsequently, I extend this spatial interest to address the spaces of puppets in popular culture that have emerged from the seven chapters, such as spatial associations, spaces of the body, interpersonal spaces, textual spaces and spaces of the border. Finally, I revisit and elaborate upon the most significant contributions arising from this work for the discipline of geography (and beyond), including a reconfiguration of biopolitics and geopolitics at an interpersonal level, advocating an alternative strand of non-representational thinking to accommodate the uncanniness of the puppet, proposing a novel approach to geocriticism, highlighting the potential for puppets to involve themselves in posthumanist debates concerning the decentring of the subject and encouraging the development of both geographies of the grotesque and geographies of puppets as distinct sub-disciplinary domains of scholarship. Ultimately, this conclusion confirms the book's opening assertion that puppets are worthy members of the geographical community and have much to offer in diverse directions.

## Research questions and answers

The research has entailed the detailed analysis of over 160 cultural works, spanning fiction, film, music and (to a lesser extent) theatre. As my interest lies in how puppets are constructed as ideas and entities in cultural contexts beyond the theatre, which is less widely considered than theatrical puppet performance/s, my emphasis has been on the reception and perception of puppets, rather than performance with puppets, and on the perspective of a

DOI: 10.4324/9781003214861-18

lay audience rather than that of puppet practitioners. My starting point was an exploration of everyday constructions of 'a puppet' in metaphorical terms in adult fiction, which immediately established a divergent view of the puppet compared to that reported by puppetry practitioners and theatre critics, but this was progressively extended, diversified and deepened as the range of cultural forms explored increased and the research focus shifted. These shifts in research focus reflected the three questions that have guided the analysis, with each question being used as an orienting device for the analyses contained in each part. This association, though, is a loose one, as although each part is guided by a specific question, the other two parts also contribute to answering that question. Consequently, the answers articulated below for each guiding question draw on work throughout the volume and not just the part to which that question was structurally but loosely attributed.

### What is a puppet?

Part 1 provided a specific but partial answer to this question, given its focus on metaphorical literary references to puppets, and each chapter generated a different construction of 'a puppet'. In Chapter 1, puppets were small in scale and dwarfed by context, with the face of a ventriloquist's dummy on the body of a marionette. It was characterised by a combination of a jerky movement and a sudden loss of movement, typically denoted as slumping, and its materiality was assumed to be wooden. By contrast, the puppet as constructed in Chapter 2 made no reference to specific types of puppet or materiality or movement, but attended much more closely and substantively to the relationship between the puppet and its master, thereby constructing the puppet as distributed between bodies rather than combined within a single body and positioning the puppet as both a reflection of and a critical commentator on society. Control is crucial to puppets – or more accurately – the lack of control is crucial to these puppets, being controlled by an external, more intelligent force with the puppet relatively infantilised, but in Chapter 1, this was focused at the level of the body through physical threat and harm, with the subjectivity of the puppetised character being evicted, whereas in Chapter 2, the control was of a more psycho-social nature involving the manipulation of situations and motivations, in which the subjectivity of the puppetised character is central. Fully consistent with the puppet as grotesque, all constructions were predicated on contradictions, incorporating, for example, flesh and wood, puppet and master, mobility and immobility, and eviction of subjectivity and evocation of empathy, thereby firmly establishing puppets as boundary beings that exist within the relations of difference between contradictory categories, in other words, the border/s.

Even in Part 1, then, we have established three distinct constructions of puppets (popular [bodily, psycho-social] and practitioner), but this was further complicated through the later chapters. In Chapter 3, puppets were found to be deep reservoirs of individual and social memory, mediators of communication (socially and psychologically) and sources of creative world-making in a series of constructions that highlighted puppets as autonomous, intelligent, agentive, controlling and often evil, thereby reversing many of the assumptions identified in Part 1. Although Chapter 4 re-established a focus on the marionette, this time the marionette lacked a face but was highly psychologised in a merging of the constructions in Part 1. This chapter also saw certain movements as definitional of puppet constructions but anchored this on two continua: (1) grounded lightness generated through exaggerated jauntiness in which a bouncy quality of movement is paired with a weightiness of tread or contact and (2) anatomical disassembly, characterised by a lack of co-ordination, spanning from the seeming disaggregation of a formerly cohesive body of parts to the disavowal of the possession of anatomical parts. Chapter 5 highlighted the diversity of puppet types but drew out the significance of faciality and pedality in establishing the puppet as a puppet through the mediation of its integration with its human sidekick, establishing that the performativity of puppets can be considered a complex mixture of excessiveness and constitutive absence. In addition, Chapter 5 suggested that puppets can be conveyors of moral messages and drew out that they are deeply and powerfully affectively charged, which was explicitly related to the uncanniness of puppets and – through that – the borderscape.

Part 3 attended to constructions of puppet transformations rather than constructions of puppets as entities, but further insights emerged from this as to diverse understandings of what a puppet is, most obviously in establishing puppets as ontologically unstable. In Chapter 6, a specific puppet – Pinocchio – was constructed as a puppet–human–animal combination, established through his materiality and sociality. Finally, Chapter 7 included constructions of puppets as vehicles for human self-actualisation, which were predicated on complementarity as much as on contradiction and integrated bodily and more-than-bodily elements and – ultimately – concluded that puppets are diverse and complex ecological–technological multicorporeal constructions.

### What grants puppets their performativity?

It would be easy to answer this question by simply referring to the contradictory (grotesque) nature of puppets, but to do so would over-simplify the matter, not least because one-and-the-same puppet can be an example of the grotesque (in combining human and puppet materialities and capabilities) but not be evidently grotesque (e.g., if the human part is hidden from view). The simple existence of contradictions and the potential for interchangeability and merging between contradictory categories is not in itself sufficient to grant puppets their performativity. Instead – and as highlighted in both Chapter 3 and the dimensional framework (Figure C2.1) – it is the central, focal and active interrogation of that contradictoriness – the puppetness – of puppets that was found to lie at the heart of the most peculiar, creative and spectacular puppets and puppet worlds. This was seen, for example, in the simultaneous eviction of subjectivity and evocation of empathy in Chapter 1, the intersection between variously conflated spaces, conjoined bodies and contrary memories in Chapter 3, the simultaneous hyper-puppetisation and de-individualisation of Mr Punch in Chapter 4 and the combination of excessive faciality and a constitutive absence of legs in Chapter 5.

There are, though, other factors that intersect with this, relating specifically to potentiality, artificiality, complementarity and sociality. The greatest potential arises when both human and puppet are liberated from the constraints of their material form and existential possibilities, and the absence of puppets as material forms in the works analysed in Chapter 2 was a limitation to the potential of these works as (most of) the plots were confined to purely human contexts and psycho-social forms of control. Revealing rather than concealing the artificiality of the puppet can be used to generate a range of effects from the comedic (*Team America*) to the confronting (*Dead Silence*) in Chapter 2, while in Chapter 5 its sudden revelation was found to prompt especially strong affective responses (*War Horse*). This was discussed at length in the context of the relation between the border and the borderscape and it becomes significant again here because the proposed spatialisation of the relationship between the explicit (cognitive) and the implicit (affective), whereby the former is focused on the border but the latter maintains the borderscape, might help to account for the fluctuation between belief and disbelief and the ability to maintain implicit belief despite the explicit knowledge of the artifice. This potential engagement with puppets on both an implicit and explicit basis, then, could hold considerable significance not only for the performativity of puppets but also for their distinctly uncanny quality.

While liberation from existential constraints and the revelation of artifice can be highly productive of peculiar puppets and puppet worlds, it remains important for audiences to be able to relate to those puppets and worlds, and it is in this context that complementarity becomes significant. In Chapter 2, complementarity was important in enabling the puppet masters to enter the social circles of their victims and in making credible (within the fictional world being created) the evolution of the victims' relationships to their puppet master, while in Chapter 7 establishing complementarity was an important feature of the transformations from puppets to humans, for example, through affinity or familiarity, again paving the path of credibility for a more substantial merging of human and puppet. Alongside this compositional complementarity sits sociality or more specifically, the socially situated nature of these puppet plots, which not only helps the reader or viewer to relate to the characters and worlds being constructed but also enables the puppet to speak back to society, either revealing or critiquing aspects of social life, as was highlighted in Chapter 2. These puppet constructions, then, are not only performative in the sense of generating new worlds but also in the sense of delivering social commentary, which – it was suggested in Chapter 5 – could be utilised more fully in relation to moral messaging.

Overall, then, puppets are at their performative peak when their own puppetness is directly interrogated through the construction of the puppet in a careful balance between optimising potential and maintaining relatability and can be elevated further by the sudden revelation of the puppet's artificiality. This can both heighten affective engagement with the puppet being (de) constructed and trigger an experience of the uncanniness of the puppet due – potentially at least – to the mutual construction of the puppet on both implicit and explicit levels. Moreover, this performativity can be extended beyond the performance setting to social reality, establishing puppet constructions, plots and worlds as both socially and geographically relevant, and positioning puppets as repositories of political potential.

### How are puppets constructed?

In this volume, puppets have been shown to be constructed in four distinct yet related ways: Through a process of borderscaping, through the combination of complementarities and contradictions, through diverse multicorporealities and in the micro-spatialities that emerge. As a starting point, puppets are constructed through myriad and diverse border-crossings between human and puppet (and sometimes animal) within a broader scope – or scape – of possibility. The relationship between the border and the borderscape has been postulated as being important to setting the parameters both for the puppet that is constructed and for the experience of the puppet as uncanny, as the implicit belief at the level of the borderscape can be sustained even as the connection forged at the border itself is broken at the explicit level. A continuum of borderscaping strategies was identified (theatrical–radical strengthening or dissolution of the border), as detailed in Figure 5.1, which can be used individually or in combination with different elements of the borderscape.

The contradictory nature of puppets has long been acknowledged and has been reinforced in this volume, but the role played by complementarities has also emerged here in two senses. On the one hand, complementarities are used to align puppet and human features to make the construction – and especially the transformation (in Chapter 7) – more credible within the narrative world being created. On the other hand, rendering congruent was identified as a strategy used by the puppet masters (in Chapter 2) to ingratiate themselves to their victims and insert themselves into their victim's social circles. While puppets are rooted in contradictions, then, these are sometimes supplemented by complementarities for specific purposes. These contradictions and complementarities gave rise to the identification and conceptualisation of diverse approaches, devices, technologies and ecologies (see especially Chapters 6 and 7), which were subsequently integrated into three key diagrams: The multicorporeality of puppet-becomings (Figure 7.2), the borderscape (Figure 7.3) and the ecologies and technologies of puppet-becomings, also referred to as the constructional framework (Figure C3.1). While each of these can be considered in isolation, they are also related as the constructional framework provides more detail on the multicorporeal devices, while the borderscape sits below the pancorporeal ontological space in setting the parameters of potential for a specific puppet rather than containing all possible permutations of bordercrossing. This borderscape, then, describes conceptually a sub-set of possible multicorporeal devices, within which an even smaller sub-set are enacted as actual border-crossings through which a given puppet is constructed.

There is, though, one final way in which the puppets considered in this volume have been found to be constructed, or more accurately, potentially undermined, and that is through the micro-spatialities that emerge from narrative and/or illustrative versions, which might or might not coincide with the underpinning borderscape and which might or might not be consistent with one another (see Chapter 6). In this instance, not only was the borderscape found to preclude the construction of the puppet purported to be constructed, by omitting the potentiality of puppet sociality, but the inconsistency between the relational spaces within the narrative and the illustrative aspects of the work (Morpurgo's *Pinocchio by Pinocchio*) was deemed to undermine the construction of the puppet protagonist. These micro-spatialities, then, can be utilised as a means of exploring and critiquing puppet constructions and narrative integrity, and can either reinforce or undo the constructive work otherwise achieved by the borderscaping, complementary and contradictory, and multicorporeal practices of puppet construction.

## Puppets and the grotesque

The puppet's association with the grotesque was acknowledged in the introduction, and numerous examples of the grotesque have been identified through the seven main chapters. My interest in this section is neither to repeat those examples nor simply to confirm that puppets are an example of the grotesque, but rather to specify the most significant implications of these analyses for our understanding of the grotesque in both conceptual and spatial terms.

Conceptually, the first way in which the grotesque was constructed somewhat unconventionally in this volume was in the consideration of popular and practitioner understandings of the puppet in Part 1, which prompted the framing of these contradictory views as an ideational body grotesque of the puppet, rather than being associated with a specific aesthetic form or material object. A second reconceptualisation was the recognition that although contradictions are at the heart of the grotesque, grotesque forms can also entail complementarities, and that those complementarities can play an important role in the combination of contradictions so central to the grotesque. This was first unearthed in Chapter 2 as a strategy of puppet masters 'rendering congruent' to insinuate themselves into their victims' social circles but was reinforced in Chapter 7 as puppet and human forms were brought into alignment to facilitate a human–puppet transformation. A third reconceptualisation proposed the notion of the character contorted as a specific formulation of the grotesque for explicitly or primarily psychological examples of the body grotesque, for example through the modification of a puppetised character's thinking. On this reading, the body grotesque and the character contorted form equivalent and paired terms. Finally, it was proposed that while we might consider the grotesque to be a concept without form (Harpham 2006), we can equally think of the grotesque as a form beyond conceptualisation due to its existence in the connections between contradictory categories, making it difficult to place in any category. Conceptually, then, there are several ways in which these grotesque puppet constructions inform and can advance our work with the grotesque, providing alternative perspectives (ideational rather than aesthetic), countervailing characteristics (complementarities as well as contradictions), additional specificity (the character contorted), and a mirror-image understanding of the grotesque's relationship to concept and form.

The last of these, in referring to the connections between conceptual categories, brings us to the spatialisation of the grotesque, as this also reconfigures the conventional notion of the grotesque existing in the gaps between (rather than the connections between) those categories (Kayser 1963; Pilný 2016). Focusing on the creative reconfiguring of borders that connect as much as they separate transforms the difference-as-distance implication of the gap between conceptual categories into an aspatial gulf of ontological im/possibility that is bridged by the same relations of difference that hold the categories apart, thereby generating new spatialities of the grotesque. On this reading, the grotesque is not best considered as topographical but as topological (or at least, as much topological as topographical), as discussed in Chapter 3, in a further reconceptualisation of the grotesque.

A further way in which the grotesque has been spatialised through these analyses is in the recognition of specific and diverse ways in which the character contorted can be constructed, over and above the merging of two identities in one body (as in supernatural puppet narratives). On the one hand, Chapter 2 identified instances in which a puppetised character's subjectivity is manipulated through the contortion of social and professional networks in a distributed or networked character contorted, but on the other hand, Chapter 4 drew out the externalisation of Mr Punch's character to the percussive background in an explicitly spatial manoeuvre. These examples also highlight how the body grotesque and the character contorted can inter-relate in different ways, as in some instances (e.g., the supernatural) the body grotesque and the character contorted might be coterminous within the body of the puppetised victim but in the case of the puppet masters in Chapter 2, the two-body image of puppet and master is distinct from the distributed construction of the character contorted, whereby the latter is larger than and encompasses the former. Consequently, while we can retain the notion of the grotesque as a space of political potential (Bakhtin 1984; Danow 1995; Li 2009; Duggan 2016), we should also be more attentive to the ways in which it can be variously spatialised, generating its own geographies.

Through these collective conceptual and spatial considerations, this volume both reinforces the need to examine the grotesque more explicitly and the value in doing so (Pilný 2016) and helps to redress the acknowledged dearth of scholarship on the grotesque itself (Connelly 2012). Puppets are an instantiation of the grotesque and can constitute a specific body grotesque, but they are also a fruitful lens through which to reconsider, reconceptualise and re-spatialise the grotesque.

## Spaces of puppets in popular culture

Looking across the diverse array of spaces and spatialities that have emerged through the foregoing chapters, it is possible to discern five spatial categories, each of which accommodates multiple spatialisations. The first of these relates to conventional spatial associations, the second concerns the puppet–human bodily form, the third relates to interpersonal and social spaces, the fourth to textual spaces and the last one concerns spaces of the border.

Conventional spatial associations of puppets include the range of spaces identified in Chapter 3 from everyday to iconic spaces, which also identified specific spaces explicitly associated with puppet origins/histories (e.g., attics, cellars) and relationships (e.g., bedrooms), while Chapter 6 identified additional specific spaces associated with Pinocchio (e.g., the fireplace, the puppet theatre). Chapter 3 also drew out the potential for spaces to be puppetised (modified for explicitly puppet purposes), which was distinguished from fantastical worlds in that the puppetised spaces are constructed through the interrogation of puppetness deemed so important to the creation of the most spectacular puppet worlds. Moreover, Chapter 5 highlighted the critical distinction between puppets and Muppets, whereby the fictional spaces are modified by Muppetisation (based on their individual characters) rather than puppetisation, as in excluding the human operators from the visible performing entity The Muppets deny their own puppetness. Puppets, then, have both their own spatial preferences and the capacity – through their puppetness – to modify and create their own spaces and places, but they cut themselves off from this ontogenetic potential if they ignore what they are.

Spaces of the conjoined human–puppet body can be distinguished between those associated with the process of becoming conjoined and those associated with the breaking down or incompleteness of that composite form. The former – conjunctive spaces – relate to processes of merging and transposition, etc., which gave rise to the continuum of four spatialities – associated respectively with object, humanised object, objectified human, human – that were generated through different combinations of aspect (bodily/psychological), focus (form, materiality, subjectivity) and process (merging, transposing, evicting, extending), as shown in Figure 1.1. They also include the spaces of memory (individual and social) from Chapter 3 and the evacuated space of subjectivity that is simultaneously and paradoxically an affective space of empathy from Chapter 1. Those associated with the breaking down or incompleteness of the composite puppet–human form – termed here disjunctive spaces – include the space of the disassembling/disassembled body identified through the analysis of mood and movement in musical constructions of puppets (Chapter 4), the constitutive absence of puppet legs that emerged in Chapter 5 as an important factor in rendering human and puppet equivalent in specific puppet–human configurations and the deconstruction of the puppet–human form by the radical strengthening of the puppet/human border through the withdrawal of the puppeteers from the 'corpse' of the puppet horse, also in Chapter 5. In each of these latter cases, the space of the puppet was perceptually and affectively maintained across a differentiated and distributed set of human and puppet elements, whereas in the former (conjunctive) spaces there existed a singular and coterminous human–puppet space, albeit that they are configured in different combinations of border-crossings.

Interpersonal spaces of the puppet can be considered at two scales: The interpersonal proper between puppet and puppet master and the social/societal. The interpersonal spatialities include the spatialising manoeuvres of incorporation, situation and disposition, whereby the puppet master established a similarity or complementarity to their victim; the spatialising orientations of deflection, reflection and rejection of the puppet master's real intentions toward their victim, generating an illusory upward trajectory, a topological non-trajectory or an actual and blatant downward trajectory for the puppetised character; and the genre-specific approaches of yoking (analogy), attaching, possessing and obsessing as means of forging relations between the puppet and its master (all in Chapter 2). The social/societal spatialities include the socially mediated, networked nature of control in distributed constructions of the character contorted in Chapter 2, the redistribution of Mr Punch's character to the percussive and choral background in his musical construction in Chapter 5 and the productivity in puppets for the purposes of critical social commentary and disciplinary impact identified in Chapter 2 and for moral messaging identified in Chapter 5. Finally, the varied social spatialities (human, puppet, animal) constructed in relation to Pinocchio in Chapter 6 were found to be especially significant due to the potential discrepancy between these constructions, which risk undermining the integrity of the puppet being constructed. Thus, the spaces generated by the construction of puppets are – or at least can be – constitutive of that construction, which renders puppets inherently geographical.

Textual spaces of puppets include those emergent micro-spatialities (e.g., Pinocchio's varied social spatialities) that can reinforce or undermine the puppet being constructed, but Chapter 6 also highlighted the narrative and illustrative as distinct yet related spaces of puppet construction, which can scaffold or contradict each other, while Chapter 7 drew further attention to the significance of intertextuality in the context of new tellings of pre-existing works and raised questions as to the extent to which an original puppet construction should determine the borderscape for a subsequent construction of the same puppet. While Chapter 7 focused on the intertextual construction of a specific puppet (Mr Punch), similar issues had emerged in Chapter 4 in relation to puppets in general, through consideration of different interpretations of Keith Harris and Orville's *I Wish I Could Fly* in the context of broader puppet discourses.

Finally, and perhaps most significantly – in uniting puppets, the grotesque and geography – are the spaces associated with the border: Topological gulfs, the borderscape and multicorporeal spaces. The border has been established as connecting as much as separating the elements on either side of it, reconfiguring it as a topological gulf (in Chapter 3), and it is through these connections that puppets are constructed; so puppets are border creatures, inhabiting and constituted in these topological gulfs. However, the array of potential topological gulfs that can be bridged in puppet construction, and the myriad ways in which they can intersect, establish an ontological field of possibility around the border/s concerned (the pancorporeal space). Within this sits the borderscape for any one puppet construction, which potentialises the puppet being constructed, which is subsequently actualised through the specific border-crossings that are employed. The process of borderscaping can employ different strategies, from theatrical to radical strengthening or dissolution of the border (Chapter 5). Interwoven with these strategies are numerous devices concerning bodily, psycho-social and contextual factors, with ten such devices being specified in Part 3 and conceptualised as intercorporeal (bodily) and epicorporeal (more than bodily) devices that might or might not be informed by pericorporeal (contextual) considerations. Some of these devices can be considered more technological (associated with puppets) and others more ecological (associated with humans), although these categories – in true grotesque fashion – are not absolute but interwoven as the pericorporeal space and the borderscape are both ecological and technological. This, in turn, introduces the significance of the relationship between the border and the borderscape, which – in constituting its own topological gulf – explodes into existential reality as the uncanniness of the puppet, whereby a disconnect between the two has been associated with a contradiction between implicit (pre-reflective or affective) and explicit (perceptual, cognitive) engagements with the puppet. Thus, the borders – or topological gulfs – that puppets inhabit are associated with and generate a suite of other spatialities that are constitutive of puppets.

Ultimately, then, puppets are associated with certain spaces, they are constructed through spatial practices, they create their own spatialities, and we can understand both their construction and their peculiar uncanniness in spatial terms. They are inherently and thoroughly spatial, making them inherently and thoroughly geographical.

## Puppets and geography

The Conclusions to Parts 1, 2 and 3 outlined specific geographical contributions arising from the analyses within each part, and this section both revisits and further develops these, with a fivefold focus on the dimensional and constructional frameworks, bodies and the more-than-bodily, specific disciplinary perspectives and sub-disciplinary domains, affect and geographies of the grotesque.

By way of a brief overview of the contributions arising from this volume, Part 1 highlighted the scaling effects of puppets speaking to social commentary and critique and the downscaling of the puppet's combination of geopolitics and biopolitics to the interpersonal level, enabling disciplinary engagement with puppets as a means of both public engagement and societal impact. It also highlighted ways in which the bodily and psychological can be both held apart in different puppet narratives and intricately interwoven in a single narrative – such as the spontaneous eviction of the subject and evocation of empathy for that subject – with implications for disciplinary efforts to decentre the subject (to which I return below). Part 2 delivered most of the conceptual and theoretical contributions in relation to both puppets and the grotesque, including the borderscape, strategies of borderscaping, topological gulfs and the uncanny, which – along with the dimensional framework (Figure C2.1) – provide fruitful avenues for further scholarship, many of which are discussed further below. Part 3 presented a supplementary constructional framework (Figure C3.1) of ecologies and technologies of puppet construction organised around a suite of multicorporeal devices and drew out an alternative understanding and practice of geocriticism attentive to the implicit spaces that emerge from a narrative rather than the explicit spaces that are written about in the narrative. One question that remains, however, is what additional benefits might be discerned if we consider these contributions collectively?

The dimensional and constructional frameworks (and associated/supplementary diagrams) provide a collection of conceptual, empirical and methodological prompts to stimulate further work not only in relation to puppets but also for exploratory application to other related cultural forms (e.g., animatronics and artificial intelligence) and other two-body or composite ideas (e.g., hybrid and socio-nature). Moreover, the dimensional framework provides an entry point for any academics wanting to engage with puppets and their fictional narratives for the purposes of communicating their own research in creative ways, by suggesting ways in which this might be done (e.g., for social critique, affective impact or communicative mediation), while the constructional framework gives pointers as to specific devices that might be employed in doing so (e.g., intercorporeal or epicorporeal devices).

Drawing heavily on the puppet's association with the grotesque has inevitably placed much emphasis on the body, which is entirely consistent with the prominence of the body in contemporary geography, but the proposal of the character contorted as a paired concept with the body grotesque puts equal emphasis on the psychological, which is inconsistent with current interest in decentring the human subject as recognising the unique cognitive capacities of the human risks re-exceptionalising the human. This, though, is where the contrariness of the puppet might prove especially helpful through its capacity to sustain simultaneously the eviction of the subject and the evocation of empathy for that absent subject. The space of the body is concurrently a vacant space of subjectivity and an affective space of empathy in a mirror image of the concurrent radical strengthening of/at the border and the continuance of the implicit/affective puppet across the borderscape. It is objectified but at the same time is re-subjectified by that very process of objectification, while that re-subjectification does not remove its objectivity, hinting at a way to accommodate a minimal humanism within a posthuman discipline (Braun 2004; Simpson 2017). While Guattari's idea of the cosmos has been proposed as a means by which to release subjectivity from the shackles of its humanist grounding by drawing on a maximum of ontological universes to go beyond the determinations of human experience (Hynes and Sharpe 2021), the puppet affords the same generativity but without jettisoning the human entirely. The puppet, then, potentially provides both a conceptual and material tool for philosophical and theoretical debates about subjectivity and a route to a posthuman humanism, and puppetivity was proposed in Part 3 to hold potential as a replacement term for subjectivity and a candidate concept for a collective subjectivity (Thrift et al. 2010), due to its capacity to decentre the human while – nonetheless – keeping it in oblique focus.

At the sub-disciplinary level, several domains of geographical scholarship are potentially advanced by the emergent findings of this volume. As outlined in Part 1, the biopolitical and geopolitical are mutually embedded in puppet constructions and both are brought down to the interpersonal level, while also enabling puppets to communicate social critique. This, then, facilitates a more integrated and smaller-scale perspective on these aspects of political geography and provides additional tools and avenues for the continuing development of critical geography. Part 2 contributed to the further development of conceptual work stemming from both political geography (the borderscape) (dell'agnese and Amilhat Szary 2015; van Houtum and Eker 2015) and comic book geography (the topological gulf) (Dittmer and Latham 2014), spoke to methodological debates about the need for proficiency or expertise in the geographies of music (Wood et al. 2007; Wood 2012) and provided an alternative perspective on the relationship between the cognitive and the affective (Banfield 2016), for more detail on which, see below. Moreover, Part 2 extended the relevance of this work to more-than-human geographies beyond the human–puppet relationship to animal sociality, materiality and concerns such as commodification and abuse, again providing new opportunities for conceptual, empirical and critical work in these areas of the discipline. Finally, Part 3 provided further contributions in articulating an alternative understanding and practice of geocriticism, potentially opening new avenues for cultural geographical engagement with fiction and further highlighting the significance of thinking about and researching through implicit as well as explicit registers of experience and knowledge, which brings us to the penultimate substantive contribution, relating to affect.

Affect was identified as an undercurrent running through the successive analyses, most notably in relation to empathy and expressiveness (or lack thereof), but it came to the forefront of the discussion in Chapter 5 in the context of the unexpected affective power of the dramatic revelation of the artifice of the puppet, which might conventionally be expected to shatter the illusion of vitality and subjectivity and thereby nullify the affective investment (belief) in the puppet. This was related to the uncanniness of the puppet whereby viewers are considered either to fluctuate between or simultaneously maintain belief and disbelief (Gross 2011; Francis 2012; Jurkowski 2013). These were integrated with the idea of the borderscape to suggest that the border is more closely associated with the explicit (cognitive awareness of the puppet's artificiality) whereas the borderscape is more associated with the implicit (pre-reflective or affective investment in the puppet). This was attributed to two factors. The first of these factors is that as the implicit investment is grounded in a multiplicity of ways in which the puppet is constructed, one or more borders can be brought into sharp cognitive focus without destroying the implicit totality of the borderscape. The second factor is that the contradictory categories that have suddenly been torn apart are proposed to remain connected on an implicit level, in accordance with a line of non-representational thinking from beyond geography. It is the latter factor that is of greatest significance here, as this alternative non-representational perspective, which allows for greater interaction between representational and non-representational thinking and specifically for some capacity to dig into our pre-reflective or affective experience cognitively (Banfield 2016), also holds broader relevance to the grotesque. There is a clear parallel to be drawn here between the implicit-explicit duality in the puppet and the simultaneity of explicit (cognitive) awareness of the impossibility of integrating the contradictions in the grotesque and implicit (pre-reflective or affective) commitment to that integration. In a Gendlinian sense, we can feel the totality of the implicit whole even though we know that it is an impossibility. Not only did the three-way integration of the uncanny, the borderscape and the implicit facilitate a spatialisation of both the implicit and explicit across the borderscape, and the uncanny across both the borderscape and the implicit/explicit distinction, but also provided an alternative non-representational perspective on the uncanny and the grotesque, as well as on the puppet. Drawing more heavily on Gendlin's philosophical and psychotherapeutic work, then, would be a valuable extension and variation of conventional non-representational geography.

One final set of disciplinary contributions relates to geography's engagement with the grotesque, which is notable for its paucity, seemingly even more so than the discipline's engagement with puppets. A search of the academic database Scopus in July 2021 for the terms 'geography' and 'grotesque' under title, keywords and abstract returned only eight works, of which only two were in explicitly geographical publications. While Scopus is clearly not infallible, as it failed to identify a geographical paper on the grotesque cited earlier in this volume, such limited identifiably geographical work on the grotesque highlights fertile conceptual and aesthetic terrain for further exploration. This is especially notable in the context of this earlier publication, which similarly advocated the expansion of geographical work on the (body) grotesque (Thorogood 2016), an expansion that is seemingly yet to happen. Thus, I echo this advocacy but also encourage progression from an emphasis on the body grotesque's vulgarity and political potential (Thorogood 2016; Brigstocke 2020) to consider the grotesque more broadly across the discipline in relation to its compositional contradictions, its topological unification of those contradictions, its ambiguous conceptualisation, its diverse formulations, its highly affective nature and its explicitly spatial aspect. In other words, I encourage progression to the scholarly exploration of grotesque geographies and the geographies of the grotesque as much as I encourage expansion of geographical engagement with puppets and the idea of the borderscape. The borderscape is as relevant to the grotesque as it is to puppets, as both are constructed on contradictions. Puppets are examples of the grotesque and can take explicit form as a body grotesque (but also the character contorted), and both puppets and the grotesque have been spatialised across implicit and explicit registers of experience, thereby accommodating the uncanniness of both the puppet and the grotesque. Puppets, the grotesque, the borderscape and the uncanny are both co-mingled and co-implicated, providing a rich resource for disciplinary excavation in forging grotesque geographies of diverse borderscapes.

## Conclusion

This conclusion opened by providing responses to the guiding research questions. The question of what a puppet is proved the trickiest to answer succinctly, as even Part 1 established divergent understandings between popular and practitioner understandings and between passing and narrative metaphorical uses. More generally, puppets have been variously defined according to size, form, materiality, sociality and mobility, and have been shown to be both contradictory and complementary in nature, and both a reservoir of personal or social memory and a communicative mediator both psychologically and socially in function. Moreover, puppets have been established as transformational in their own right as well as vehicles for human self-actualisation, sources of affective potency and vehicles for social commentary. A puppet, then, is many things.

The other two research questions were somewhat easier to summarise. In terms of what grants puppets their performativity, puppets were found to be at their performative peak when their own puppetness is directly interrogated in a careful balance between optimising performative potential and maintaining relatability and can be elevated further by the sudden revelation of the puppet's artificiality. Further, this performativity can be extended beyond the performance setting to social reality, establishing puppet constructions, plots and worlds as both socially and geographically relevant, and positioning puppets as repositories of political potential. With respect to how puppets are constructed, puppets have been shown to be constructed in four distinct yet related ways: Through a process of borderscaping, through the combination of complementarities and contradictions, through diverse multicorporealities and in the micro-spatialities that emerge from the borderscape/s from which they spring, with numerous and substantial overlaps between these four mechanisms.

Puppets are undeniably an instantiation of the grotesque and can constitute a specific body grotesque, but they are also a fruitful lens through which to reconsider, reconceptualise and re-spatialise the grotesque, as this volume has shown. Conceptually, the popular and practitioner understandings of the puppet have been proposed to instantiate the grotesque as ideational as much as aesthetic and material in nature, while complementarities as much as contradictions have been identified as constituting the grotesque, with the former being a strategy through which the combination of the latter is sometimes achieved. The character contorted has been proposed as a supplementary term for psycho-social forms of the body grotesque, while recognising that they might overlap and interact in diverse ways, and the grotesque has been framed as a form without concept as much as it is a concept without form.

Puppet spatialities were then considered, and five spatial categories were discerned, the first of which relates to conventional spatial associations, the second concerns the space of the puppet–human body, another evokes interpersonal and social spaces, the fourth relates to textual spaces and the last one concerns spaces of the border. It has been shown that everyday spaces can be puppetised or Muppetised, and a distinction has been drawn between conjunctive and disjunctive spaces of the puppet–human body in which the former are coterminous, but the latter are differentiated. A further distinction has been drawn between the interpersonal spatialising manoeuvres of the puppet master and the social (not societal) distribution of puppet–puppet master constructions, although the capacity for puppets to function as critical social commentators extends this spatiality to the societal level. In addition, both textual and intertextual spaces were identified as important to the construction of puppets, and border spatialities emerged as multiple, varied, inter-related and important in the construction of puppets. The borderscape was proposed to potentialise the puppet, which is subsequently actualised (or not) through the specific borderscaping practices employed, which – in turn – can be considered in terms of either strategic approaches to the border (theatrical–radical strengthening or dissolution of the border) or specific multicorporeal devices (ecologies–technologies). Moreover, the relationship between the border and the borderscape was found not only to be important but to be (potentially at least) significant to the uncanniness of the puppet through the construction of the puppet on both cognitive (reflective) and affective (pre-reflective) registers. Spatialising the pre-reflective engagement with the puppet across the borderscape and the reflective engagement with the puppet at the border facilitated the integration of the borderscape, the uncanny and the cognitive-affective nexus.

Finally, the disciplinary contributions arising from these collective endeavours were elaborated, including the potential within the dimensional and constructional frameworks to stimulate and guide future geographical work with, on and through puppets, and the specific sub-disciplinary contributions with respect to: Political geography (the integration of biopolitics and geopolitics at the interpersonal scale), cultural geography (the conceptual development of the topological gulf and related ideas in comic-book geographies and an alternative – implicit – approach to geocriticism) and more-than-human geographies (human–puppet–animal sociality, the potential for puppets to both convey moral messages and deliver affectively impactful moments, and their capacity to utilise those moments to drive home the moral message).

More significantly, though, are the contributions arising with respect to non-representational and posthuman geographies and geographies of the grotesque. An alternative strand of non-representational thinking from beyond geography – Eugene Gendlin's philosophical and psycho-therapeutic work – was introduced as a way of accommodating perceived shortcomings in dominant non-representational understandings of the relationship between the cognitive and the affective with respect to accounting for the uncanniness of the puppet. This was deemed to provide an underpinning perspective that can unify the borderscape and the uncanny puppet, through notions of explicit and implicit understanding that relate to each other more bidirectionally than conventional non-representational geographical ideas of affect. Secondly, the contrariness of the puppet was deemed to hold productive potential in the context of contemporary debates over the status of the human and the nature of subjectivity. In this context, the simultaneous evacuation of the subject and evocation of empathy for that subject, and the inherently more-than/less-than-human multiple-singular nature of the puppet was suggested as providing a new conceptual and material form and practice through which to engage with these issues. Specifically, it was suggested that puppetivity might provide a replacement term for subjectivity and provide a conceptual candidate to encapsulate collective subjectivities, wherein puppets can – in this context (but not a performative context) – function very productively as proxies for humans. Taking this narrative and argument full circle, this suggestion was grounded in the puppet's capacity to dissolve and accentuate borders simultaneously as this means that we can decentre but not entirely jettison the human through the theatrical dissolution of the human/puppet border, which keeps that border (and thereby human distinctiveness) both out of sight and in focus at the same time. Finally, geographical underengagement with the grotesque as much as with puppets was identified and the co-implication of the puppet, the grotesque, the borderscape and the uncanny was highlighted as providing fertile terrain for the exploration of grotesque geographies of diverse borderscapes and the establishment of an explicit strand of geographical scholarship on the grotesque alongside the development of the geographies of puppets and puppetry.

The chapters in this volume have delivered a thorough and wide-ranging (although not exhaustive) exploration of how puppets are constructed in popular culture from the perspective of a lay spectator to supplement existing work on puppet performance from the perspective of the practitioner. Despite its strong analytical focus on representations, this volume has engaged in detail with the affective and disconcerting (uncanny) power of the puppet, thereby emphasising the non-representational quality and performativity of representations, and has forged strong links between geography and the grotesque with potential benefits in each direction. These chapters have generated multiple contributions – empirical, conceptual, methodological and theoretical – for scholarship on puppets, on the grotesque and in geography (and beyond), by – for example – considering the construction of the puppet and the function of the uncanny on both explicit and implicit registers (encouraging theoretical diversification in non-representational geography), reconfiguring our understanding of the grotesque (on spatial grounds), downscaling the intersection of geopolitics and biopolitics to the interpersonal level and instantiating both the geographies of puppets and the geographies of the grotesque.

In multiple ways, then, puppets are long overdue recognition as inherently spatial entities and should be capturing distinct and sustained disciplinary interest, as should the grotesque. Puppets live and are constituted in the topological gulf, they puppetise the spaces with which they are associated, they generate their own unique spatialities (e.g., bodily, interpersonal), they re-spatialise the (body) grotesque, they are moulded or undermined by the implicit micro-spatialities of the works in which they can be found, and they are both born of and rendered uncanny by the borderscape. Puppets fully deserve to take their place as valid and valuable members of the geography community.

Bring on the puppets!

## References

Bakhtin, M. (1984) *Rabelais and his world.* Indiana University Press: Bloomington, IN.

Banfield, J. (2016) *Geography meets Gendlin: An exploration of disciplinary potential through artistic practice.* Palgrave Macmillan: New York.

Braun, B. (2004) Querying posthumanisms. *Geoforum*, 35: 269–273.

Brigstocke, J. (2020) Resisting with authority? Anarchist laughter and the violence of truth. *Social and Cultural Geography.* doi:10.1080/14649365.2020.1727555.

Connelly, F.S. (2012) *The grotesque in Western art and culture: The image at play.* Cambridge University Press: New York.

Danow, D. (1995) *The spirit of carnival: Magical realism and the grotesque.* The University Press of Kentucky: Lexington, KY.

dell'agnese, E. & Amilhat Szary, A.L. (2015) Borderscapes: From border landscapes to border aesthetics. *Geopolitics*, 20: 4–13.

Dittmer, J. & Latham, A. (2014) The rut and the gutter: Space and time in graphic narrative. *Cultural Geographies*, 22: 427–444.

Duggan, R. (2016) *The grotesque in contemporary British fiction.* Manchester University Press: Manchester.

Francis, P. (2012) *Puppetry: A reader in theatre practice.* Palgrave Macmillan: Basingstoke.

Gross, K. (2011) *Puppet: An essay on uncanny life.* University of Chicago Press: Chicago, IL.

Harpham, G. (2006) *On the grotesque: Strategies of contradiction in art and literature.* The Davies Group Publishers: Aurora, CO.

van Houtum, H. & Eker, M. (2015) Borderscapes: Redesigning the borderline. *Territorio*, 72: 101–108.

Hynes, M. & Sharpe, S. (2021) Cosmic subjectivity: Guattari and the production of subjective cartographies. *Area*, 53(2): 311–317.

Jurkowski, H. (2013) *Aspects of puppet theatre*, 2nd edition. Palgrave Macmillan: Basingstoke.

Kayser, W. (1963) *The grotesque in art and literature*. Indiana University Press: Bloomington, IN.

Li, M.O. (2009) *Ambiguous bodies: Reading the grotesque in Japanese Setsuwa Tales*. Stanford University Press: Stanford, CA.

Pilný, O. (2016) *The grotesque in contemporary Anglophone drama*. Palgrave Macmillan: London.

Simpson, P. (2017) Spacing the subject: Thinking subjectivity after non-representational theory. *Geography Compass*, 11: e12347.

Thorogood, J. (2016) Satire and geopolitics: Vulgarity, ambiguity and the body grotesque in *South Park*. *Geopolitics*, 21: 215–235.

Thrift, N., Harrison, P. & Anderson, B. (2010) "The 27th Letter": An interview with Nigel Thrift. In: B. Anderson & P. Harrison (eds) *Taking place: Non-representational theories and geography*. Ashgate Publishing Ltd: Farnham and Burlington, 183–198.

Wood, N. (2012) Playing with 'Scottishness': Musical performance, non-representational thinking and the 'doings' of national identity. *Cultural Geographies*, 19(2): 195–215.

Wood, N., Duffy, M. & Smith, S.J. (2007) The art of doing (geographies of) music. *Environment and Planning D: Society and Space*, 25(5): 867–889.

# Index

Note: Page numbers in *italics* denote figures, page numbers in **bold** denote tables

Printed in the United States
by Baker & Taylor Publisher Services

Printed in the United States
by Baker & Taylor Publisher Services